MW00340024

MILES AND MILES OF TEXAS

MILES AND MILES OF TEXAS

100 Years of the Texas Highway Department

Carol Dawson with **Roger Allen Polson**

Geoff Appold Photo Editor   Foreword by **Willie Nelson**

TEXAS A&M UNIVERSITY PRESS | COLLEGE STATION

Copyright © 2016 by Highway Centennial Book, Inc
All rights reserved
Second printing, 2017

This paper meets the requirements of ANSI/NISO Z39.48–1992 (Permanence of Paper).
Binding materials have been chosen for durability.
Manufactured in China by Everbest Printing through FCI Print Group

Library of Congress Cataloging-in-Publication Data

Names: Dawson, Carol, 1951– author. | Polson, Roger Allen, author.
Title: Miles and miles of Texas : 100 years of the Texas Highway Department /
    Carol Dawson and Roger Polson ; Geoff Appold, photo editor ; foreword by
    Willie Nelson.
Description: First edition. | College Station : Texas A&M University Press,
    [2016] | Includes bibliographical references and index.
Identifiers: LCCN 2016015905| ISBN 9781623494568 (cloth : alk. paper) | ISBN
    9781623494575 (e-book)
Subjects: LCSH: Texas. Department of Transportation—History. | Texas.
    Highway Department—History. | Texas. State Department of Highways and
    Public Transportation—History. | Highway departments—Texas—History. |
    Highway engineering—Political aspects—Texas—History. | Texas—Politics
    and government—20th century. | Ferguson, James Edward, 1871–1944.
Classification: LCC TE24.T4 D39 2016 | DDC 388.109764—dc23 LC record available at https://lccn
    .loc.gov/2016015905

In the beginning was the road.—Joseph Bedier

Contents

Foreword

by Willie Nelson

Listen to my song, and if you want to sing along, it's about where I belong, Texas.

Sometimes far into the night, and until the morning light, I pray with all my might to be in Texas.

I am a guitar picker from Abbott, Texas. I always have been, and always will be, proud of the state where I grew up. Since the 1930s I have had the opportunity to watch Texas grow with me. Small footpaths that cut through cow pastures or cotton farms slowly turned into gravel roads that turned into highways that eventually turned into interstates. Over the years, my bands and I have driven a million miles playing music from Amarillo to Brownsville, El Paso to Nacogdoches, and from Austin to the rest of the world. These highways have taken me far, but I always come home to Texas.

So as you are driving through our beautiful state, wave if you see me. I am out there *on the road again* somewhere.

It's where I want to be, the only place for me. Where my spirit can be free, Texas.

MILES AND MILES OF TEXAS

1

Tracks, Trails, Mud, and Money

Highways are the foundation of our economy, of our society, of what we are.—Ray Barnhart, administrator (retired), Federal Highway Administration

The Texas Highway Department was born in corruption.—Marcus Yancey, former deputy executive director, TxDOT

At the precise moment on April 4, 1917, when Governor James Ferguson touched his pen to paper to sign the law creating the Texas Highway Department, he ignited a battle between the forces for the public good and the evil of greed. The worst offender was Jim Ferguson himself. Second was his wife, Miriam, who also became the first woman in America elected to a state governor's seat in a general election and who acted as the Ferguson figurehead after her husband was impeached and banned forever from holding a Texas office. Others included their cronies and hangers-on, who, like the Fergusons, saw the brand-new Texas Highway Department as a field fresh and ready plowed for their personal plunder.

The combat would last fifteen years. Its warriors would practice their strategies of advance and retreat, attack and defend, each side struggling to maintain power and control over the prize: the largest, wealthiest department or agency ever devised for any state in the United States. And when it finally ended, that same department would leap forward and grow to become, at least until the new millennium, one of the country's cleanest, most innovative, most efficient instruments of progress.

As this street scene in San Benito demonstrates, by 1911 the automobile was rapidly overtaking the horse and buggy in every corner of Texas. (Courtesy of the Library of Congress)

This is the story of how the war raged and how its outcome has affected the state of Texas, the nation, and the world. Of course, the story is about much more than the clashes between a thieving pair of politicians and a few good people. It is a tale as vast as Texas and covers not only 268,820 square miles of landscape but also 350 years of "bad road" history, as well as the last 100 years of changes in Texan culture, community, industry, agriculture, and individual lives. It is a picture of extremes, as is so much about the Lone Star State: spectacular weather, drastic road conditions, exceptional stubbornness, dishonesty, and integrity. Most of all, it is the story of people determined to build a superior road system—and a new civilization—almost from scratch.

The Birth of Texas Roads

Thousands of years before *Homo sapiens* commenced their treks across the expanses of the North American continent, a maze of trails already threaded the landscape. Migratory animals moving through grazing grounds in search of food, salt licks, and water sources as the seasons changed were the first pioneers to tread the paths into place. Three of the early species to institute effective routes across rough terrain were mastodons, mammoths, and musk oxen. Later the bison and deer followed, retracing the safest steps between bogs and deserts, finding the easiest passes through which to cross mountain ranges, ford rivers, and navigate prairies—inscribing the earth with their passage.

This was particularly true of the region that would become Texas. All of its territory except the southernmost tip of land arcing along the Gulf of Mexico hosted the Plains bison. According to some estimates, in 1492, the year Christopher Columbus arrived in the New World, the number of bison roving across most of North America was approximately 60 million. Although other experts dispute this theory, one fact is certain: during the first decades of the nineteenth century, tens of millions of the animals throve and fed on the grasslands created by American Indian burn-offs—by far the largest population of a single species of mammal on earth. By several eyewitness accounts, a person could stand fixed in one spot on a migration route and watch as a single herd in constant motion took three days to pass by. Only the regions now designated as Washington, Arizona, Nevada, the majority of California, and the edge of the Eastern Seaboard remained untouched by the enormous groups.

Nomadic Plains peoples followed these trails to hunt, confirming the routes as the most accessible, with the best supplies of water and food. They also used them for trade, as had the eastern-based Mound Builders for three to five millennia before them. Thus, for the European settlers who later ventured westward, the hope of acquiring a home where the buffalo

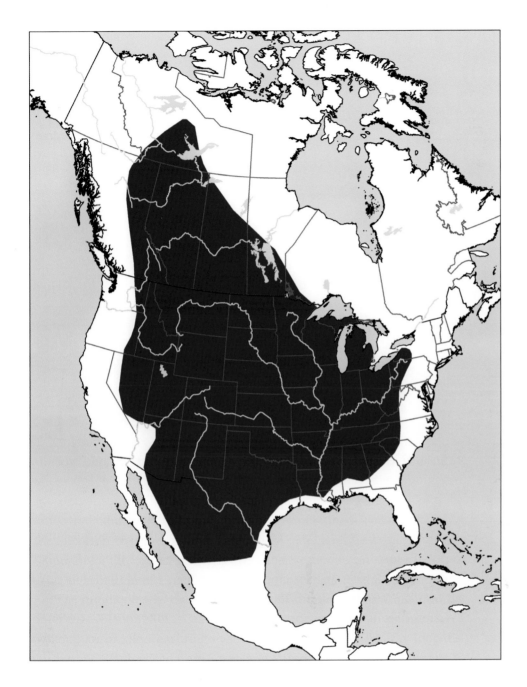

Historic bison range for North America. (Courtesy of NatureServe)

roamed was more than mere whimsy; it was a practical necessity, since these sites had already been proven by thousands of years of travel.

The hunting tracks that developed from the game traces encircled the outer edges of future Texas, leaving the middle open and relatively unmarked by comparison, and ensuring eventual connections for healthy long-distance trade that reached from South and Central America all the way up to the Michigan area and beyond. Later, these same trailways would provide a blueprint for the Spanish to overlay with their web of Texan *caminos*.

The evolution of a road is an ancient and ongoing story, a continuum not just through space but through time and human change. Walls crumble;

Indian marker trees are found throughout much of the United States. They were the first "road signs." This pecan tree was bent by the Comanche to mark the California Crossing, an important low-water crossing on the Trinity River. This is a Comanche Marker Tree specific to the Comanche Nation. Marker trees were bent to guide travelers to significant locations such as campsites, water sources, river crossings, and other important natural features. Today, the tree grows in front of a National Guard building where a tank watches over it. (Courtesy of Steve Houser, Comanche Marker Trees of Texas, *Texas A&M University Press*)

empires evaporate; temples subside into jungle and dust; palaces, cemeteries, and skyscrapers are stripped and dismantled and their grounds usurped for other projects. But roads, once imprinted on the world, remain the most permanent of human endeavors. Only through road systems and the links they facilitate can societies thrive, prosper, and progress. Mobility, imperative for public construction, architecture, agriculture, urban expansion, speedy communication, marketing, and religious propagation, depends on efficiency; without it, none of these amenities would exist. Civilizations sponsor roads, and roads, in turn, promote civilizations.

They also enable sweeping victories. When the first bands of conquistadors sailed from Spain and Cuba and landed in the New World in 1519, they encountered cultures already steeped in thousands of years of advancement. It soon grew apparent that these new invaders would also find the strategy of availing themselves of previously formatted trade and trans-jungle religious routes opportune for their *entradas*, or expeditions of discovery and military aggression. Such routes proved tailor-made for vanquishing the road builders themselves. The Spanish conquered the Aztecs and the Mayans, appropriating the beautifully engineered stone and stucco-paved roads, called *sacbeob*, that the Mayans had built through the

This Mayan sacbe
was constructed
prior to 1000 AD at
the ceremonial site,
Labna, on the Yucatan
Peninsula. (Photo by
Dr. Antonio Rafael de
la Cova)

Sacbeob (singular, *sacbe*: translated into English as "white road") were marvels of labor and engineering. Elevated anywhere from 2 to 8 feet above the jungle floors of the Yucatán Peninsula, built for foot and animal traffic, their paved roadbeds consisted of walls of large dressed stones, layered more roughly in between with fill of smaller rocks surmounted by smaller ones and finished with a pale limestone stucco topping. The sides were probably mortared together and plastered with this same substance, though no remnants of it have endured. The widths of these roads could stretch 15 feet; the longest one known covered more than 186 miles. Stone rollers flattened the cement of the top surfaces and created a smooth, clean-swept causeway through many miles of dense and vine-hung foliage. Because the Mayans had no large pack animals or live-stock brawny enough to pull a cart or wagon, the wheel was not employed as a component of transportation on these roads. But its principle was well known, as demonstrated in the road-smoothing roller, and its technology appeared throughout Mesoamerica in miniature, in the form of wheeled jaguars' and dogs' feet in children's toys. Therefore, the *sacbeob* were confined to pedestrian traffic. It is generally agreed among the early Spanish chroniclers and later scholars that the dozens of *sacbeob* still traceable that ran between sacred places, temples, and ceremonial centers, although also used for trade and communication, were chiefly intended for religious purposes. Large processions of pilgrims and celebrants walked them from Coba to Yaxuna, from contemporary Mérida to Tulum.

Like the Indian trade routes to the north, like the Olmec trails that in turn became Aztec roads, like the three thousand–year-old Incan Great Road network that curves like a spine for 14,292 miles down the Pacific Coast of South America ("the most expansive piece of infrastructure relating to transportation in the New World," according to one expert), parts of the *sacbeob* were later destined to be appropriated by newcomers—first by the Spanish for their own exploratory objectives, and then by modern-day populations incorporating them into automotive highways and railway lines. Even after they had vanished from immediate sight, they still served as the basis for more contemporary thoroughfares. For example, when the United Fruit Company started building a railroad in Guatemala in 1910, workers dug a series of drainage ditches close to the tracks, the course of which ran near the outskirts of the pre-Hispanic Mayan city of Quiriguá. Only 3.28 feet down, they struck rock. They then proceeded to plumb the thick tropical humus and alluvial deposits that had long buried the whole region, excavating a splendid causeway of cut stone leading to an unknown point somewhere in the northeast.

This sacbe, *photographed by the 1934 Villa Rojas expedition in the Yucatán, was among the impressive highways built by American Indian cultures such as the Maya in Mexico and the Ancestral Puebloans of the Southwest. (Dumbarton Oaks Research Library and Collection)*

jungles of the Yucatán and the roads across mountain passes that descended to the region of Mexico City, raying outward from there in every compass direction.

After anchoring his eleven ships in 1519 at the island of Cozumel, just off the shore of the Yucatán Peninsula, Hernán Cortés, the conquistador who conquered Mexico in the name of King Charles of Castile, first learned of the rich and powerful Aztec Empire flourishing deep in the heart of the mainland. Sailing off for some reorganization and reinforcements, Cortés returned to his original landing spot at Cempoala and consulted with its Totonac chief, who advised him to take the trade-and-tribute road eastward to Xalapa and Xixo. From there he should ascend the Sierra Madre Oriental, the country's eastern mountain range, and forge onward to Central Mexico.

For this reason, the road from Veracruz to Xalapa can, in a sense, be called the road of initial conquest: the precursor or parent of all the roads to come, built by Spanish hands for wheeled vehicles, pack animals, and horses on top of the earlier foot-traffic thoroughfares. Without it, the movements of Cortés's large heavily armed, steel-armored force (more than six hundred infantry soldiers, fifteen horsemen, and fifteen cannons), heretofore unknown in Mesoamerica, would not have been able to make its way through the jungles and forests, or climb the mountains to reach the pass and the high plateau, or pursue a clear, less arduous (and more cost-effective) route to their destination. Without it, the future mule-drawn *carretas* and oxen-pulled *carros* roadbeds branching northward toward the silver mines of Zacatecas and the colonial seat of Saltillo would not have been developed. From that primary taproot would eventually spread the arteries of Spanish rule, including the *caminos reales* that veined upward through Zacatecas, to Monclova and Guerrero and Presidio, forded the Rio Grande into Texas, then further to San Antonio, San Marcos, San Saba, Waco, Abilene, Amarillo, and eastward to Nacogdoches and beyond to Louisiana.

The Spanish roads were much cruder than the Mayan *sacbeob* or even, it is speculated, the earlier Olmec versions. They resembled the medieval roads of Spain (and indeed those of much of Europe) and consisted of no greater engineering elaboration than earthen cart tracks marked along the sides with rocks, compacted by the wheels of ore-laden *carros* and oxen hooves. This continued to be true as the Spanish trekked closer to the river separating the regions of future Texas from Mexico, developing tracks leading up into the land that would become Spanish-ruled Texas. Then they traversed that land with a network of roads on top of the old game and trade trails, christening them Los Caminos Reales, or the King's Roads. These dirt-based arteries ran diagonally from the Rio Grande to a few miles beyond the Louisiana border, branching here and there along the way. Altogether the roads and their tributaries were known as the Camino Real de los Tejas.

There was little of the royal treatment for travelers on the Camino Real (King's Road) in Texas. This remnant of the trail can still be seen at Mission Tejas State Park near Crockett. (Steven Gonzales/El Camino Real de los Tejas National Historic Trail Association)

The Spanish *Camino* Systems

I n 1686 the viceroy of New Spain charged Alonso de León to lead an expedition north to quell the French interlopers led by René-Robert Cavalier, Sieur de la Salle. In his hunt for evidence of French presence, he followed the Indian trails that he thought would prove worthy of further investigation for prospective *caminos reales* and marked them with crosses slashed into tree trunks. Only a few years after his sojourns, in the early eighteenth century, he was succeeded by teams of missionaries, surveyors, military officers, and Indian laborers who staked those same routes as highways and then edged their dug-out roads with boulders and rocks. Upon its completion in 1759, the roughed-out Camino del Norte linked the Rio Grande Valley along 650 miles of road-constructed landscape, all the way to the banks of the Red River. On the other side of the Rio Grande, it continued southward to the capital of New Spain.

When Spanish ambitions fixed on Texas as a territory for religious conversion, the first Catholic outpost, Mission San Francisco de la Espada, was instituted in East Texas in 1690 and became one of the chief destinations on the early iteration of El Camino Real de los Tejas. This northeastern leg of the Camino Real eventually continued toward Nacogdoches, where another mission, Mission Nuestra Señora de Guadalupe de los Nacogdoches, named after the Caddo who lived there, was established in 1716 and terminated in Los Adaes, the original capital of the province of Texas. It also then spiked toward the French settlement of Natchitoches.

Both these outposts were placed in what is now Louisiana. During the Spanish Colonial period, El Camino de los Tejas, its branches, side routes, and loop curves, constituted the sole primary overland corridor from the Rio Grande to the Red River Valley, linking the chain of missions, posts, trade centers, and *presidios*, or forts, which the Spanish built to defend their territories from the French threat.

Although Mission San Francisco de la Espada, after a collapse, restoration, and three name changes, was eventually relocated to the hub of missions in San Antonio de Bexar in 1731, it nonetheless represented the future of Spanish intentions in East Texas, along with its sister mission founded 6 miles away only three months later: Santísimo Nombre de María. Neither of these initial East Texas mission sites survived its first two years, but simultaneous to the reconstruction and renaming of San Francisco de la Espada in 1716, the Catholic Church dedicated a third mission, San José de los Nazonis, 2.5 miles north of what is now the town of Cushing in Nacogdoches County. The local populations these missionaries hoped to convert were the Hasinai and Neches peoples, descendants of the Caddoan Mississippian Mound Builders. State Highway 21 currently incorporates that particular section of the Camino Real de los Tejas, following the same route the Spanish used, and before them, the Caddoan people and their ancestors, going back to at least 850 CE—nearly twelve hundred years ago.

A number of other campsites and outposts the *caminos* connected lay on the same route from Coahuila, which included major stations in San Antonio de Bexar, New Braunfels, San Marcos, and Austin. By the time the French threat ended with the signing of the Treaty of Fontainebleau in 1762 and the expensive, hard-to-supply East Texas missions and presidios were closed or moved as a result, the Christian conversion rate among the Caddoan groups had proven comparatively small. Los Adaes was abandoned in 1773 as irrelevant to the new political circumstances. The missions in San Antonio supplanted all the others in vitality. But El Camino de los Tejas remained an essential route for trade and migration of all kinds and continued to be so for another century. The Camino Arriba, later known as the Old San Antonio Road, that tracked through latter-day Bastrop and Crockett, now assumed an even greater importance as the route between San Antonio de Bexar and Natchitoches, Louisiana, the old

State Highway 21 currently incorporates a section of the Camino Real de los Tejas, following the same route as the Spanish, and before them, the Caddoan people and their ancestors. The Caddo Mounds, located in Cherokee County near present-day Alto, testify to American Indian settlements in East Texas. (Kevin Stillman/TxDOT)

French trade center ceded to Spain by the French under the Fontainebleau Treaty. At least two of the missions, one on that route and one on the segment known as the Laredo Road, not only played significant roles in future Texas history as sites of events during the Texas Revolution, but also served as spurs for consequential road development: La Bahía at Goliad and Mission San Antonio de Valero, later renamed the Alamo.

Goliad was the place where, in 1749, the Spanish positioned both Mission Nuestra Señora Espíritu Santo de Zúñiga and Presidio Santa María del Loreto de la Bahía, and where, nearly ninety years later, Santa Anna's army massacred 342 captured Texas Revolutionary troops outside and within the stone fortress walls. Today, US 59 (a section of the NAFTA Corridor Highway System and soon to become Interstate 69W) traverses Goliad on its way between the Mexican border at Laredo and the Canadian border in northernmost Minnesota.

The First Mail Service

As a tool of civilization, few if any factors hold more value than swift and reliable communication. Although Indian attack, weather, and other circumstances challenged couriers and rendered messages difficult to transmit, a steady mail service was deemed essential to the Spanish colonizers of Texas and did much toward furthering the settlement of the region.

The first official postal service, inaugurated in 1779, made full use of both the Upper and Lower Roads of the Camino Real network. A monthly run might carry letters on a three-month-long, 1,200-mile journey, originating at Arispe, Sonora; moving upward via Mission San Juan Bautista to San Antonio; then to New Braunfels, San Marcos, Alto, Nacogdoches, and from there to San Augustine. The Lower Road dropped down from San Antonio to Cuero before once again progressing in a northeasterly direction to rejoin the Old San Antonio Road.

In 1786, the first ferry service across the San Antonio River helped shorten the time and distance. Starting in 1792, weekly runs delivered the mail on its long journey north, aided by the construction of the military-inspired Middle Road (Camino de la Pita) between Presidio de Rio Grande and San Antonio in 1807.

By 1827, the Camino Real de los Tejas was nearly three hundred years old. Age, usage, and weather had taken their toll on what had been a rough, unkempt roadway in the first place, and the Mexican *ayuntamiento*, or legislative council, that oversaw Texas governance insisted on improvements to the mail and stagecoach routes, recognizing that these were also militarily necessary for ease of troop movement if Mexico was to continue to enforce its authority.

The first peaceful contingent of American Anglos to defy Spanish law and slip into Texas reached the jutting finger of land on the Red River that the French called Pointe aux Peconques, or Pecan Point, in the summer of 1811. Long before the French recorded its geographical utility as a campsite, it had been occupied by Caddoan people named Natchitoch. Unlike previous interlopers, the "filibusters" who had entered Texas to stir trouble and try to found an empire, this particular pack of Americans (about a dozen in number) nourished no aspirations to create a republic or independent state or to declare themselves rulers or kings. After traveling on an ancient buffalo trail across the Red River into Northeast Texas, the group secreted themselves on this peninsula, where they constructed a small shelter and lookout, remaining until sometime before June 1815. The fact that they were outlaws on the run no doubt necessitated extra stealth and caution.

After their departure, the next newcomers had more reputable aims: two former army supplier brothers named George and Alex Wetmore erected a trading post alongside the old buffalo river crossing. Soon a competitor opened shop nearby, and the Anglo-American wave of settlement began in earnest. Walter Pool and Charles Burkham were the initial arrivals, followed shortly thereafter by Claiborne Wright. Pecan Point, a center of community rather than an actual town, marked the first truly permanent Anglo-American habitation in Texas.

Although it is perhaps not surprising that the earliest transient Anglo residents of this fresh frontier were fugitives from justice, the precedent seems clear: Texas was to gain a reputation as a last-chance outpost, a refuge for criminals, debt dodgers, and misfits, among many citizens throughout the United States, and the malefactors' ingress now looks as much prophetic as symbolic. This trend was later borne out by the "Gone to Texas" slogan written on the house doors of abandoned Southern homesteads after the Civil War. Such escapes were often as much flights from authority as they were quests for a new start in life, and perhaps inflected the Texas attitude toward government, dictates of any kind, stubborn independence, and self-determinism for many years to come. Certainly such stances affected the future attempts to form a state government department dedicated to improving the Texas roads.

After the *ayuntamiento*'s Decree of April 6, 1820, stating that Mexico required passports of any foreigners to enter Texas, Anglo-Americans attempting to evade detection by Mexico blazed a number of "sneak traces" through the thick, dense pinewoods of East Texas. These did not constitute actual roads, of course, but were crude trails bristling with hacked stumps, sodden and soft with pine needle mats, sticky from red clay mud after rainstorms, swamped in sloughs and bayous, rife with mosquitoes, and patrolled by dangerous snakes and other animals. All these were merely some of the perils presented to travelers venturing upon them, especially if the users tried to negotiate them in a wagon. One such example became known as the Tennesseans' Road due to the number of Anglo immigrants from that state who used it. Meanwhile, the poor conditions of El Camino Real de los Tejas itself, as well as its branches, hindered any effective efforts by the Mexican army to curtail these trespassers, with decayed tracks vanishing into gullies and ravines, river crossings that proved almost impossible to ford, or the merest trails overgrown by scrub that had to be chopped apart with machetes. Despite the *ayuntamiento*'s desires, few of the projected road improvements were ever realized.

Then another element arrived to underscore the lack of easy egress and compound the need for better roads. After gaining permission and land grants from the Mexican government, Moses Austin and his son Stephen F. Austin started an Anglo-American colony in Texas.

New Travelers

The Austin settlers entered Texas by both land and sea. But inevitably they made their way to their new holdings via the road systems developed by the Spanish and first traced by buffalo herds and the American Indians who had hunted them. When Stephen F. Austin mapped the roads available through Texas, he elaborated in detail the first routes used primarily by the Anglo travelers encouraged by Moses Austin. More "roads" and traces were blazed by these same settlers trying to reach their land grants. Repeatedly, the records of muddy roads cited by travelers lend a sharp picture of what travel was like.

The worst travel nightmare, though, arrived in March 1836 during the Texas Revolution, when people tried to evade the Mexican army's deadly bayonets in a dash for safety that became known as the Runaway Scrape.

It started with a message of defeat. When news reached settlements that General Santa Anna had overpowered the American and Texian forces at the Alamo and was advancing through the countryside, most of the Texian settlers cringed in fear. After hearing of the massacre Santa Anna had ordered for the Alamo survivors, they could only imagine what grisly punishment he planned for them.

Desperate to escape the Mexican army, hundreds of families dropped everything and fled their homes with only the clothes they were wearing, in their panic often leaving even the necessities of food and water behind. There were few roads to take, and those were soon jammed with refugees. Men, women, children, babies: they trudged through the rain and mud; they gathered at riverbanks in throngs, stranded by floodwaters; they died by the hundreds. The numerous eyewitness accounts provide not only the glimpse of environmental circumstances in a moment of crisis but also a general description of the challenges to be met at almost any time, depending on weather, over the next several decades by those wishing to move over Texas territory.

Many women and children as well as men died in the efforts to traverse the rugged, trackless terrain and flooding rivers across the lower half of Texas. They were buried on the spot where they fell. Estimates of casualties range from dozens to hundreds of settlers; in all, the military losses in the Revolution added to loss of civilian lives in the Runaway Scrape equal between 10 and 20 percent of the Texian population at the time.

However, the poor conditions of the existing roads also provided an advantage to the revolutionaries that same month. The cold, wet spring besetting the region made infantry passage difficult. Mexican troops slogging through East Texas mud and the rutted, trenched roadbeds of the Camino Real to Bexar (San Antonio) and then eastward found the route exhausting, which ultimately played a role in the defeat of the Mexican army. General

Stephen F. Austin, often called the Father of Texas for his irreplaceable leadership of early colonists, drew the first accurate map of Texas to assist other Americans who wanted to immigrate to the then-Mexican territory. (Courtesy of Texas State Library and Archives Commission)

In the words of one Runaway Scrape participant, Dilue Rose Harris, the daughter of a doctor who kept a careful journal of the events, the terrors of trying to outrun the Mexican army were escalated by tragedy:

Every one was trying to cross first, and it was almost a riot. We got over the third day, and after traveling a few miles came to a big prairie. It was about twelve miles further to the next timber and water, and some of our party wanted to camp; but others said that the Trinity river was rising, and if we delayed we might not get across. So we hurried on. When we got about half across the prairie Uncle Ned's wagon bogged. The negro men driving the carts tried to go around the big wagon one at a time until the four carts were fast in the mud. . . . Our hardships began at the Trinity. The river was rising and there was a struggle to see who should cross first. Measles, sore eyes, whooping cough, and every other disease that man, woman or child is heir to broke out among us. . . . The horrors of crossing the Trinity are beyond my power to describe. One of my little sisters was very sick, and the ferryman said that those families that had sick children should cross first. When our party got to the boat the water broke over the banks above where we were and ran around us. We were several hours surrounded by water. Our family was the last to get to the boat. We left more than five hundred people on the west bank. Driftwood covered the water as far as we could see. The sick child was in convulsions. It required eight men to manage the boat. When we landed the lowlands were under water, and everybody was rushing for the prairie. Father had a good horse, and Mrs. Dyer let mother have her horse and saddle. Father carried the sick child, and sister and I rode behind mother. She carried father's gun and the little babe. All we carried with us was what clothes we were wearing at the time. The night was very dark. We crossed a bridge that was under water. As soon as we crossed, a man with a cart and oxen drove on the bridge, and it broke down, drowning the oxen. That prevented the people from crossing, as the bridge was over a slough that looked like a river. Father and mother hurried on, and we got to the prairie and found a great many families camped there. . . . The other families stayed all night in the bottom without fire or anything to eat, and with the water up in the carts. The men drove the horses and oxen to the prairies, and the women, sick children, and negroes were left in the bottom. The old negro man, Uncle Ned, was left in charge. He put the white women and children in his wagon. It was large and had a canvas cover. The negro women and their children he put in the carts. Then he guarded the whole party until morning. It was impossible for the men to return to their families. They spent the night making a raft by torch-light. As the camps were near a grove of pine timber, there was no trouble about lights. It was a night of terror. Father and the men worked some distance from the camp cutting down timber to make the raft. It had to be put together in the water. We were in great anxiety about the people that were left in the bottom; we didn't know but they would be drowned, or killed by panthers, alligators, or bears.

Santa Anna, confident that he had cornered the rebel Texians, finally allowed the weary foot soldiers a two-day rest stop at San Jacinto, a decision fatal to his presupposed victory. When General Sam Houston's army descended on them during the afternoon siesta hour, the Mexican troops recovering from their long, arduous march lay sleeping, unaware and ill prepared, and in less than twenty minutes, Houston's forces won both the battle and the future republic's independence.

Frontier Forts, Civil War, and American Indian Suppression

From the establishment and legitimization of the Republic of Texas onward, a steady influx of settlers secured the prosperity of the developing territory. The focus grew on the Austin, Houston, Galveston (the two most important seaports), and the Northeast Texas communities along the Red River, leaving Nacogdoches to rusticate in its pinewood forests and San Antonio to molder as a quaint but obsolete colonial Spanish outpost. Trade began to boom. Cotton farmers and ranchers could at last sell their produce at free-trade market prices rather than Spanish- and Mexico-controlled, presidio-enforced rates. Exports flowed from the ports to New Orleans and other destinations. Although the Republic's treasury was impoverished, the economy flourished. But due to lack of access over much of the huge territory, the main population of Anglo settlers remained confined to the river bottoms of East and South Texas and along the curve of the Gulf Coast.

The 1846 annexation of Texas into the United States precipitated the Mexican-American War, which lasted until 1848 and ended with the Treaty of Guadalupe Hidalgo transferring ownership of the American Southwest and California to the United States. This made the US government responsible for protecting both sides of the Texan-Mexican border, including the Hispanic people still living, as they had for hundreds of years, on either side of the Rio Grande. Such an obligation presented a grave challenge. The Kiowa and Comanche remained an ongoing threat, not only to most of the rural Texas frontiers but to these old Spanish haciendas, villages, and towns along the Rio Grande that they had raided since their original founding. The empty westernmost reaches of the new state were even more daunting to defend, and far more dangerous for the intrepid Anglo pioneers who staked out their new homesteads than they had been for the Spanish explorers who never attempted to cultivate the arid plains. Although the Texas Rangers established a number of "forts," these had no heavy ordnance such as cannons and cannonballs or stout ramparts and walls like the stone presidios of the Spanish.

Then, in 1849, the federal government began a plan to safeguard all the pan-Texas mail, stagecoach, and wagon routes—in particular, the route to

California—with a series of forts manned by both US infantry and cavalry planted along the Rio Grande in a line from Brownsville to Eagle Pass and northward to the Red River. The military surveyed additional potential roads, which included the vital Lower Military Road, later to be known as the San Antonio–El Paso Road, and defined the passable and navigable network for travelers. These new military roads also coincided with the California Gold Rush, which saw heavy traffic generated through the wildernesses of the Chihuahuan and Sonoran Deserts by the hopeful prospectors making their way to the gold fields. In 1851–52, the US Army began further construction— this time a line of forts about 200 miles west of the previous line, which soon assumed even more strategic importance than the first set of defenses, with a few exceptions, such as Fort Clark at Brackettville.

The Civil War and Texas' secession from the Union and allegiance to the Confederacy put an end to the US government forts' occupational functions. The Federal troops and officers were evicted after the initial firing on yet another fort, far away in South Carolina: Fort Sumter. Not until the war ended in 1865 would security efforts on the frontier resume with any efficacy, and then under very different circumstances, involving a different foe. Blood-thirsty bandits, defiant Mescalero Apache, and other American Indians opposed a multiethnic complement of white soldiers, Black Seminole scouts (the descendants of slaves escaped from Southern plantations into the sanctuary of Seminole tribes inhabiting the Florida Everglades), and Buffalo Soldiers—the African American cavalry troops who manned isolated Fort Davis and fought the same Plains Indians who had given them their nickname. Patrolling these dry, dust-ridden regions on a hardy cavalry horse, or even a camel (as was instituted in 1856 when the US Camel Corps brought the beasts to Camp Verde in Kerr County and later on the route to Fort Davis), was a grueling chore. Trying to push through the washboard hardpan, spiked scrub, and rocky sand in a buckboard, a Conestoga wagon, a stagecoach, or an entire wagon train or pack-mule convoy was a feat. Fort Lancaster, located in the Pecos River Valley only a short distance from the present-day, spacious 80-mile-per-hour Interstate 10, was another lonely station manned by Buffalo Soldiers. It had been built to shield the mail carriers— three coaches per month—as well as the immigrants, military supplies, and freight consignments that moved along the San Antonio–El Paso Road. (Today's I-10 is, in fact, essentially the same as the old San Antonio–El Paso Road, with a few brief deviations here and there due to modern construction and cut-throughs.) Fort Lancaster also became a stopover for the Camel Corps on its journey westward. Established in 1855, the fort was initially staffed by infantry and did not see its first African American contingent until 1867. Late that same year, in December, the forty officers and Buffalo Soldiers occupying the fort fought off a Kickapoo horse-stealing party of approximately four hundred warriors. Only three soldiers died in the skirmish. Thus, Fort

Lancaster, despite the twentieth-century multitude of western films and television scripts depicting otherwise, became the sole Texas army post ever to be directly attacked by Indians.

During the Civil War, the Confederacy appropriated several of the forts, along with the eighty camels left behind by the US troops who had been permitted to surrender their posts, gather up their equipment, and return to the northern Union. The Confederates continued the attempts to guard the San Antonio–El Paso Road and the surrounding frontier, but eventually the decision was made to desert the far western forts such as Lancaster, Stockton, and Davis as too burdensome. Not until the war ended were the forts and the road-protection and mail-escort services revived by the US Army.

Throughout the conflict, the Confederacy relied on Texas as a crop provisions state. Texas ports also served as an alternative entry for supplies from Europe to reach the South, and the Southern-Pacific Railroad terminus established before the war in Marshall, in East Texas, became a strong link in the resource chain, although it stretched no farther into the region beyond Marshall's close proximity to the Louisiana border and the depot in Shreveport, Louisiana. But the bad roads and dilapidated amenities such as abandoned forts along them, as well as Indian attacks farther west, hampered any other useful role the state could play. They also fettered travel: during the war, all stagecoach service west of San Antonio halted completely, except for the line connecting San Antonio to Eagle Pass. Further ingress into the western regions—long nominated as Indian Hunting Grounds and Territory through treaty between Republic president Sam Houston and the Waco, Delaware, Chickasaw, Tawakoni, Kichai, Caddo, Anadarko, Ionio, Biloxi, and Cherokee—was simply too hazardous.

Texas Travel in the Mid- to Late Nineteenth Century

Soon after the Civil War ended, the Reconstruction period commenced. The majority of Texans resented this policy with its federal government oversight as fiercely as did many residents of the other Southern states. Its legacy influenced—and poisoned—Texas attitudes and politics for decades to come, affecting even the progress of road building in the state.

During the Reconstruction years, from 1865 to 1877, the governor's seat was filled by so-called Radical Republicans Elisha Pease and Edmund Davis. In 1871, Davis demanded that each adult male in the state register for work duty on road, bridge, and state-sponsored public school crews or else pay a one-dollar poll tax, or "road tax." This tax was repealed in 1874, when the Anti-Republican Redeemer Democrats prevailed to regain control of Texas politics and Democrat Richard Coke was elected governor. But it inaugurated a rash of new legislation addressing the roads predicament.

The heyday of the cattle drives was brief. From 1866 to 1886, millions of cattle were rounded up and driven north to the railhead in Kansas. The Chisholm Trail was about 50 miles wide in some areas from the immensity of the traffic. (Courtesy of Texas State Library and Archives Commission)

Manchaca Springs near Onion Creek in present-day Austin was a known stopping place for travelers as far back as the days of the Camino Real. The Weir family operated the stagecoach stop here from the 1850s until the arrival of the railroad in 1881. (Courtesy of Texas State Library and Archives Commission)

Many of the long-distance routes through Texas over the next few years of the 1860s and 1870s were developed by cattle drives and bullock-drawn freight carriers—most especially, the track of the future I-35—following those same ancient trails first created by animals and Paleolithic hunters. These routes, however, failed to help most of the common people who might wish to leave their farms or ranches, their towns and cities, or move goods without punitive expense. They found it difficult to go anywhere in relative efficiency or comfort, unless they happened to live in East Texas or near a riverbank with a steamboat wharf. An ox cart hauling three bales of cotton might traverse only a few miles a day, and the standard price of wagon transport was

A ferry operated at the confluence of the Comal and Guadalupe Rivers near New Braunfels from the 1840s until the railroad arrived in 1880. (Courtesy of Texas State Library and Archives Commission)

FERRY, COMAL RIVER.

20 cents per ton-mile. In the 1850s, as later, stagecoach passenger fares usually ran to 10 cents per mile (equal to roughly $3 today), unless rain-swollen streams and muddy roads caused the cost to double. Inadequate drainage, nonexistent rural bridges, costly conveyance, no road continuity, rainstorms, northers, and other chill and searing weather: all these factors rendered trans-Texas travel almost impossible and kept the populace stationary in comparison to that in other states. Due to the scarcity of snowfall, even the chance of a swift winter sleigh ride was not an option. Nothing about this changed right after the Civil War. "Gone to Texas" declarations were all very well; the problem was, how could anyone roam around it once they got there?

And then came the railroads.

Iron Wheels, Steel Tracks

In 1854, a certain traveler complained that after ten days of rain, it had taken him more than thirty-six hours to ride by stagecoach from Houston to the town of Hockley, 36 miles away. In 1857, that very same passenger recorded that he boarded a railroad car on the Houston and Texas Central going the same distance and arrived at his destination in one hour and forty minutes.

After the Civil War, the Houston and Great Northern Railroad Company was the first new railroad to start laying building tracks, reaching Palestine in September 1872. Meanwhile, in 1867 the Houston and Texas Central

Railroad resumed the construction that the war had interrupted, trending northward at a steady clip, reaching Corsicana in 1871, Dallas in 1872, and the Red River in 1873. It also directed a Western Division toward Washington County and finished in Austin on Christmas Day 1871. From across the state's northern boundaries, the Missouri, Kansas and Texas Railway Company forged toward Denison, achieving it on Christmas Eve 1872. Once the Houston and Texas Central reached Denison from the south the following year, a historic moment took place: the Texas railroad joined the entire national rail network in a connection that promised to end the degree of rugged isolation that had previously been both the state's mystique and its curse.

In 1870, the total railway track mileage in the state stood at 700. By 1880 the number had quadrupled to 3,293. By 1890 the railroad companies were publishing extensive Texas maps, with track lines, depots, termini, towns, villages and small communities, suggested accommodations at various stops that catered to passenger trade, whistle stops and crossings where passengers and freight could board, and the rivers and creeks winding through

The railroad transformed more than Texas transportation—it brought with it modern life, including electricity, the telephone, and the telegraph. The Texas State Railroad, founded in 1881 to service the ironworks at the state penitentiary, now operates as a living history in the Piney Woods between Rusk and Palestine. (Stan A. Williams/TxDOT)

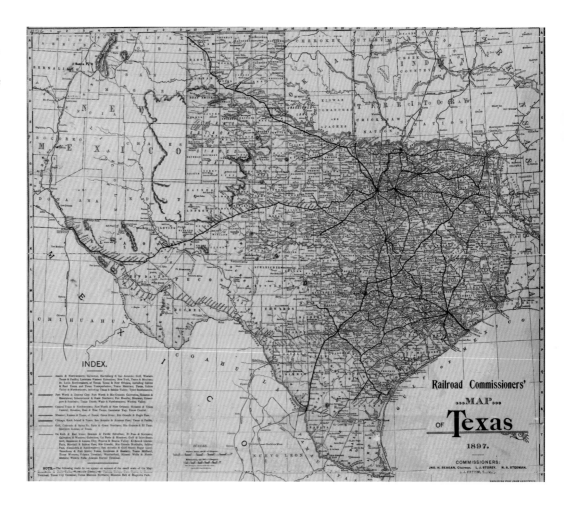

Railroad construction in Texas took off in the 1870s, leading to the rise of "King Cotton." (Courtesy of Texas State Library and Archives Commission)

Texas and Pacific Station, Fort Worth, 1909. (Special Collections, University of Houston Libraries)

the topography. The trickle of goods and people grew to a flow, and then a torrent, as more and more track was laid. But the rural dwellers who constituted the largest percentage of the population, the ranchers and farmers, still found themselves marooned on their property, defeated by distance, locked in by bad weather, unless they owned stout vehicles and horses, mules, or oxen strong enough to plod through the mud or freezes. Townspeople were little better off, unless they lived in a community with a rail depot or crossing. Getting anywhere beyond home often still presented an intimidating struggle.

Roadworthy Ambitions

In tandem with rail developments and the massive cattle drives of the 1870s, the road systems stirred increasing attention. Since the establishment of the Republic, the real responsibility of roads in Texas had rested with the military or with local community activity. Although a Commission of Roads had been authorized in 1836 to expand and control roads, toll bridges, and ferries, there had been no money in the Treasury to accomplish these aims. In 1850 the county courts were authorized to follow the old French and English corvée system model of delegating road work to counties and communities and levying labor from the pool of resident able-bodied males, requiring all between the ages of fifteen and fifty to spend up to six days a year each, laying out and maintaining local roadways—a sketchy and inconsistent procedure at best.

The historic Waco Suspension Bridge was completed in 1870 as a means to bring cattle across the Brazos River on the Chisholm Trail, predating by nearly fifty years the formation of the Texas Highway Department. It remains a legendary icon of downtown Waco and visible from present-day I-35. At the time of its completion, it was the longest single-span suspension bridge west of the Mississippi. The bridge was built with cable supplied by the John Roebling Co., who built the Brooklyn Bridge in New York City. This postcard depicts the bridge in 1911. (Special Collections, University of Houston Libraries)

Suspension Bridge, Waco, Texas.

From the early twentieth century, the bridge served vehicle traffic until 1971, at which time it became reserved for pedestrians and special events. It was listed in the National Register of Historic Places in 1970 and received a state historic marker in 1976. (Jack Lewis/TxDOT)

Incumbent upon the county commissioners' courts were the tasks of appointing a five-person commission to deem what kind of road was necessary, where, and to whose benefit, and then cutting trees and protruding stumps (which, if less than 8 inches in diameter, must be sawn off, and if larger, rounded so that wagon wheels could pass over them), clearing a 40-foot-wide path, leveling it as smoothly as possible with shovels and rakes (always with the knowledge that the first wagon wheels after the next hard rain would groove the surface in deep, water-filled ruts), and then regulating the road within their own bailiwicks, using their local citizens for all the labor. Those who preferred not to contribute their own muscles and elbow grease to the job had to produce replacements—usually in the persons of slaves or, after the Civil War, either hired hands or convicts. Sometimes these work-party gatherings of men and boys turned into another kind of party altogether, with games, music, gossip, poker rounds, and even sack races. A few fistfights broke out from time to time as well.

With the exception of the German communities in Comal County, at New Braunfels, Fredericksburg, and Fort Mason, which hired professional roadmasters to come in, design the roads, and oversee the local men building them, the results were predictable and often lamentable. The usual beneficiaries of selection for road creation by the counties' road commissions were landowners—the larger, richer, and more influential, the better. These felt entitled to good, dependable, comfortable mobility to and from their properties, both for themselves and for their harvests, while rural dwellers, tenant farmers, small-parcel holders, and the like suffered their secluded exiles without simple access to the ordinary social and economic advantages known to city residents—including good schools, libraries, stores, churches, railroads, paved streets, and trolleys. Unfortunately for everyone, this scheme stayed fixed for at least the next seventy-five years.

But concerns were escalating. For example, the Preston Trail, later known as the Old Preston Road, which had originally been developed for military purposes, ran from Preston on the Red River through Dallas to Austin. Its maintenance, though, was subject to motivation on a county-by-county basis. Therefore, any uniformity or steadiness of upkeep remained unlikely. This proved hard on the stagecoaches and merchants, to say the least. Bridges throughout the state were still few and far between. Ferries and foot-fordings continued as the primary means to cross most rivers. Another, more effective system had to be found to fund, promote, and maintain Texas roads and their components.

Because Reconstruction had scrambled or dismantled a number of Texas mandates, in 1879 the Texas legislature passed a law once again authorizing counties to lay out, supervise, and maintain roads. At that time the law revised the age limits, stipulating instead that all males between the ages of eighteen and forty-five be co-opted to work on roads for up to ten days per annum. A constitutional amendment in 1883 gave counties the right to levy a road tax of 15 cents on every $100 valuation of property. In 1887, the legislature permitted counties to issue bonds for bridges. In 1890 a second amendment was adopted, adding another 15-cent tax on every $100 valuation. All vehicle registration was run by the counties, and the county courts collected the proceeds. But still no valuable structure evolved to bring the condition of the thoroughfares beyond the most primitive, dust-flumed, mud-choked pathways. Then, from an unexpected and entirely novel source, a voice of protest lifted through the nation, campaigning for all states everywhere to improve their roads.

"Texas Good Roads"

In the 1880s, the bicycle was a brand-new invention. Early in its public career it had taken the form of the velocipede, or "penny-farthing," with the front wheel standing high as an average man's chest or shoulder, a smaller rear wheel, and an elevated seat that rendered balance precarious. The later version, the "safety bicycle," lowered both the wheels and the seat to find a more stable center of gravity and incorporated gears and a chain mechanism to assist propulsion. This version became popular in the mid-1880s and wildly so in the 1890s, triggering the bicycle craze of the Gilded Age, which would eventually affect or even initiate several different modes of transportation, including the invention of the internal-combustion automobile engine and the first practicable airplane (after the Wright Brothers converted their bicycle repair shop to an aviation engineer's dream factory). Thousands of people, both men and women, jumped onto their new vehicles and gripped the handlebars with joy, experiencing a mobility and ease of navigation formerly limited to horseback and completely denied to the standard available buggy. The bicycle meant sudden freedom. Unless, of course, the riders wished to leave the city limits and travel cross-country.

In 1880, a large group of two-wheeled enthusiasts and manufacturers, already members of individual bicycle clubs, merged their organizations to form the League of American Wheelmen (LAW). In May of that same year, the same group, under the new league's aegis, founded the Good Roads Movement. Not only did this association quickly grow to become the foremost of its kind, providing assistance to its members; recommending or endorsing products, equipment, and services; holding conventions; and encouraging

The Good Roads Movement promoted better roads for farmers to move their crops to railheads, while the League of American Wheelmen promoted the latest transportation sensation, the bicycle. An 1895 federal study prompted by lobbying from both groups found that Texas had the worst road conditions in the United States. (Courtesy of Hugh Hemphill, Texas Transportation Museum)

camaraderie; its offshoot movement also tackled the job of altering legislation on federal, state, and local levels. In 1891, it published a pamphlet that would continue to sway readers across the continent: *The Gospel of Good Roads: A Letter to the American Farmer.* This tract argued that for farmers, improved roads meant reduced transportation costs (fewer horses hauled the same tonnage), much time saved in traveling back and forth to town and to neighbors' houses, and far fewer damaged wagons and lamed horses. The farmers joined with the LAW and then the railroads to shape a new lobbying group: the National League for Good Roads. As vocal persuaders, the wheelmen and their allies made a powerful force—so powerful a constituency, in fact, that their agitation for smoother, more improved roads became a national political faction and prompted Congress to appropriate $10,000 to launch a road inquiry sponsored by the federal Department of Agriculture. The inquiry started in 1893 through the medium of the newly opened US Office of Road Inquiry, forerunner to the Federal Highway Administration. Its initial efforts remained limited to educational, technical, and promotional activities, however, because the real administration of road and highway improvements was still relegated to the counties within the states and stopped at each of those boundary lines.

The first issue of *Good Roads Magazine*, published in 1892, was decidedly bicycle-centric. Within three years its circulation reached 1 million—a strong signifier of the bicycle's viability and the fresh liberties it conferred. Before long, though, the focus of the Good Roads Movement shifted to include not only bicycles but all types of vehicular movement. Over the next few years the magazine would demonstrate the swing in political and economic interests; during that period, the movement swelled and extended, backed by journalists, farmers, politicians, engineers, and Rural Free Delivery (RFD) advocates everywhere. Partly as a result of the movement, the State of New Jersey established the first official state highway department in 1893. That same year, RFD of the US mail service commenced.

To make RFD a workable scheme, better roads were of course a necessity. The mail-order catalogue industry exerted its commercial attractions through Sears Roebuck and Montgomery Ward, inspiring farmers to urge their political representatives even harder. The railroad companies, which had rapidly embraced the National League ideas for their own purposes—chiefly the increased access to and delivery of freight—began scheduling "Good Roads" trains to rural stations. A company paid skilled workmen to bring in horse-drawn dump wagons, graders, and a steamroller for packing down the strata of stones and gravel that constituted a macadamized highway. The local people gathered round to marvel as a "paved" road gradually appeared, layered with large, melon-sized stones first, then another layer of apple-sized stones, then a third layer the size of small plums, stretching from the depot to a spot a mile or two away.

Stephen Samuel Barton was a mail carrier in 1910 in East Texas. (The East Texas Research Center, R. W. Steen Library, Stephen F. Austin State University, Nacogdoches, Texas)

Once the road was finished, the farmers made the discoveries that fueled their delight: a two-horse team could now pull the same wagon over a macadamized road (named after its Scottish inventor, John Loudon MacAdam) that had needed a four- or even a six-horse team to pull it down a dirt road. Steel wagon wheel rims and steel horseshoes compacted the stones even more firmly with use, and the traffic moved faster, with less trouble for heavier loads. Transportation costs dropped accordingly. Mud problems became a thing of the past. The roads drained rainfall from their surfaces more readily than older forms, including the bumpy, parallel-log "corduroy" roads that had been a feature in the northern United States since Colonial times. Along these fine new roads, both efficiency and crop production increased; railroads carried more produce from those same rural depots to the cities than they ever had before. The macadamized road seemed a miracle of ingenuity, the saving grace for bicyclists, farmers, postal carriers, and railroad companies alike. And it had taken these European descendants only a thousand years to finally replicate the ancient Mayan *sacbeob* technology and construction.

Unfortunately, not everyone could enjoy the macadamized road's luxuries. Many rural counties were unable to afford the backbreaking pick-and-shovel labor, the load haulage of stones, and the steamroller and other machinery required. For those who had no means to invest in or create paved roads, the improvement of existing dirt roads became a priority. During mud seasons, a horse-and-buggy rig might sink up to the axle hubs and the horses' shoulders or even necks; yet grading these quagmires without a specialized implement was fruitless, as drainage could not occur unless the road top could be "crowned," with a slope on either side. How to accomplish this without money? Crude technologies mirrored the crude raw materials. Foremost of these technologies was the split-log drag.

The King road (split log) drag, an early innovation in road building. An advertisement for the new product stated: "Easy to make — easy to run — and which, rightly used, convinces the unconvinced, converts the unconverted, makes rough roads smooth, and soft roads hard. A simple implement made on the farm, which will transform the roads of the corn and grass belt." (Illustration by Carol Dawson)

The split-log drag, also known as the King road drag or Missouri drag, was invented in 1904 by a Missouri farmer to smooth and level the soil of road surfaces. It literally consisted of a log split in half, the two sides lined up parallel with flat edges forward, fastened together 3 feet apart with three rods, and pulled by a triangulated chain or rope hitched to a mule or ox yoke. Its inventor, D. Ward King, recognizing its simplicity, did not attempt to patent the early version; instead, he traveled around the country, teaching its use and passing out handbills with step-by-step instructions. It soon became a mainstay of county road crews in Texas, as well as other regions. In some states, the government even passed laws requiring that it be part of the local equipage. Although this new invention filled a niche and served its term of efficacy, that term did not last long into the twentieth century. When Gottlieb Daimler invented the high-speed gasoline-powered internal-combustion engine in 1885, the world changed once again—at first slowly, but soon with an acceleration unequaled since the Industrial Revolution.

Daimler's invention signaled the genesis of yet another new age in transportation. It inspired and facilitated the design of an innovative vehicle: the automobile. It also inaugurated a new industry that in turn ushered in a spread of new technologies in all directions: petroleum. In addition, it at first spelled the doom of both macadamized and dirt roads, through the sheer destructive power of early automobile tires and their tendency to rip up the smoothest dirt and grind and dig ruts into the hardest-packed macadam. The ruts channeled puddles and filled with rainwater that eroded roads. In winter, the ruts then often froze into rigid peaks and combs that could in turn tear a buggy or shake a wagon to pieces. Once automobile speeds increased to 15 miles per hour, however, automobiles ceased to wreak this carnage and instead rolled swiftly from town to town on the macadamized roads.

William Faulkner

"There were already two more automobiles in Jefferson that summer [of 1905]; it was as though the automobiles themselves were beating the roads smooth long before the money they represented would begin to compel smoother roads.

"Twenty-five years from now there won't be a road in the county you can't drive an automobile on in any weather," Grandfather said.

"Won't that cost a lot of money, Papa?" Mother said.

"It will cost a great deal of money," Grandfather said. "The road builders will issue bonds. The bank will buy them."

"Our bank?" Mother said. "Buy bonds for automobiles?"

"Yes," said Grandfather. "We will buy them."

"But what about us?—I mean Maury?"

"He will still be in the livery business," Grandfather said. "He will just have a new name for it. Priest's Garage, maybe, or the Priest Motor Company. People will pay any price for motion. They will even work for it. Look at bicycles. Look at Boon. We don't know why."

— *The Reivers*, William Faulkner, 1962

Early bridges were wide enough to accommodate the old cars, but just barely. (TxDOT Photo)

Houston's Harrisburg Road, 1910. (Special Collections, University of Houston Libraries)

In 1895, the City of Houston hosted a National Good Roads Association convention, attended by those who wished to compare the experiences of different states with the problems and solutions facing Texas. The chief substance of the discussion seemed to revolve around the compulsory labor draft in the counties and the need to replace the draft with convict labor. But what emerged most memorably from the convention was a reproach from General Roy Stone, head of the US Office of Road Inquiry, who declared that Texas had made "*less progress in road development than any other state.*"

The Adventure of Colonel Ned

Soon Texas was to have its first glimpse of the gasoline-powered automobile and its first test of motoring road inadequacy, through the lens of the landmark drive of Colonel Ned Green. Edward Howland Robinson Green, born to perhaps the most famous, and also infamous, woman in nineteenth-century America, was in his way an automotive pioneer. His mother, Hetty Green, had inherited a huge fortune from her whaling family antecedents, invested it shrewdly, and caused it to multiply many times over through her foresight, risks, and cautions, eventually amassing a net worth equal to about $4.33 billion in today's valuation, which made her one of the most successful financiers in the United States or elsewhere at that time. She obsessively defended her millions, preventing her more modestly wealthy husband from touching any of her capital, contesting an aunt's legitimate will with a forgery that named her as heir, and requiring

In 1899, Colonel Ned Green imported the first car to Texas. Shown here with his driver Jesse Illingsworth, Green raced the car through the streets of Terrell at a terrifying 15 miles per hour. (Courtesy of Dallas Historical Society. Used by permission)

her future son-in-law to sign relinquishment papers to any inheritance of her daughter's money before consenting to their marriage. Her daughter was then in her thirties.

Hetty's miserliness also presented quite a spectacle: she was notorious for wearing only a single black dress and set of undergarments until they fell apart; never washing her hands or heating her shabby, unfurnished boardinghouse rooms or using hot water or soap on her person; cooking on a single-burner gas hot plate or subsisting on pies that cost 15 cents; and most tragically, attempting to have her son Ned admitted to a free clinic for indigent patients when as an eight-year-old child he broke his leg in a street accident with a handcart. After the clinic administrators recognized her and refused Ned treatment, she sought out other doctors and applied cheaper home remedies. The leg failed to properly heal; finally it had to be amputated 8 inches above the knee due to the threat of gangrene and replaced with a prosthetic cork limb. During her lifetime she received many sobriquets, among them "The Richest Woman in America" and "The Witch of Wall Street." By the time of her death at eighty-one, she was arguably the wealthiest woman in the world.

Perhaps not unexpectedly, Ned Green's extravagances seemed as flamboyant as his mother's stingy economies were demeaning. He cultivated several interests for which he became internationally noted: postage stamp collections; coin collections; automobile collections; pornography collections; prostitute harems; a collection of palatial country estates; a gold, diamond-encrusted chamber pot. An astute businessman himself, he sustained a son's loyalty to Hetty and in 1893 at the age of twenty-five took over the general managership of one of her smaller properties, the foreclosed Texas-Midland Railroad, a section of the Houston and Texas Central. At her request he moved from New York to Terrell, Texas, to facilitate turning the company with its 51 miles of rusting track into a profitable model railroad boasting "the first electrically-lit coaches in the state." As newly appointed company president, he arrived in Terrell, limped straight through the doors of the American National Bank, and handed in a $500,000 cashier's check that represented twice the bank's capital. The disbelieving bankers wired Hetty, asking about the possibility of imposture. She wired back, telling them to inspect the mole on his forehead and his cork right leg. They did so, deposited the check, and promptly made him the bank's vice president.

Although always eager to fill his leisure time by entertaining a steady stream of courtesans from Dallas and Terrell in his hotel room, Ned Green imported more than his personal tastes to Texas; he also brought with him a great deal of cash and a sense of adventure and enjoyed a new level of freedom previously unknown to his experience. Transplantation suited him. He liked the people and the freewheeling Texan atmosphere, joined the boards of several banks, throve financially and socially, and got involved in state politics. Eventually, in 1910, despite his strong Republican leanings, he supported the Democratic governor Oscar Branch Colquitt, who, after his election, conferred on Green the honorary title of lieutenant colonel of his staff in acknowledgment of his business and community contributions. Colquitt, the immediate predecessor of Governor James Ferguson, went on to serve a second term. Meanwhile, Green remained a resident for only a brief time longer before returning to New York and Massachusetts, where his mother lived under his mansion roof until her death in 1916. He sold the Texas-Midland Railroad to Southern Pacific in 1928. He therefore never encountered the full flavor of James Ferguson's political machinations and was spared the vista of future corruption in his adopted home, looming above that new frontier that he himself would be the first to breach.

Later Ned Green allegedly described those early days in the Lone Star State to listeners: "I felt wonderful. I was fancy free." Then, glancing down at his false limb, he added, "You might say I was footloose." Unlike his mother, he cherished the unhoardable: a sense of humor. For the rest of his life, unless saddened by loss or bad news, he wore a perpetual half-smile.

While still in Texas, though, his biggest road adventure occurred not in the realm of political campaigning, not while riding the rails, but on the highway, in a brand-new, two-cylinder, six-horsepower vehicle known as a gas surrey, designed by George P. Dorris and manufactured by the Saint Louis Gas Car Company—the very first gasoline-powered automobile to enter the state. And it was Green's ambition to launch the first car trip across the Texas landscape inside it.

He started out from Terrell on the morning of October 5, 1899—a Thursday—accompanied by George Dorris of the Saint Louis company. Their 30-mile route toward Dallas took them through the town of Forney, where they encountered the most serious of several mishaps: the state's first and rather inglorious auto accident, when a farm wagon crowded the gas surrey off into a ditch. The vehicle's water tank received the greatest damage, seemingly sabotaging their expedition. But a local blacksmith, an African American named Reeves Henry, was able to repair it, which in turn made him the state's premier auto mechanic. Soon the intrepid Green set off again. Five hours after their initial departure, perched high on the two-cylinder gas surrey, the pair reached the end of their journey, rolling into the outskirts of Dallas on their inflatable tires, having made a jaunt not merely through the countryside but through history.

The following day, October 6, 1899, the *Dallas Morning News* printed Green's account of the excursion:

It was amusing to notice the reaction our appearance caused along the road. Cotton pickers dropped their sacks and ran wildly to the fences to see the strange sight. And the interest was shared by the farm animals, too. One razor-back sow that caught sight of us is running yet, I know. At least a dozen horses executed fancy waltz steps on their hind legs as we sped by and but for the fact we went so soon out of sight, there would have been several first-class runaways. We did not put on full power on the country roads because it would have been too dusty for comfort.

Less than three months later, the century turned. Soon other horseless carriages, both experimental, such as homemade versions of the ethanol-fueled Ford Quadricycle, and of regular production, such as the Stanley Steamer and the Duryea Motor Wagon, began to jounce or toddle along the city streets of Texas and even attempt to explore pastoral scenery as Green had. In a little over ten years after the first issue of *Good Roads Magazine*, the bicycle had almost faded from the pages, as well as from the intent of the Good Roads Movement itself. Automobiles now dominated road development. The special-interest champions of the Good Road Movement had switched their sights and their political pressures from the obsolete concerns of the hobbyist and sportsman to the future. And cars, whether steam driven, electric powered, or gasoline propelled, were now in Texas to stay.

The Skirmishes: "Get the Farmer out of the Mud"

The Good Roads Movement bore fruit nationwide. By 1910, twenty-nine states had their own functional highway departments, commissions, agencies, or authorities dedicated to guiding, managing, and increasing the quality of the networks within their borders. By 1913, almost all of the forty-eight states had followed suit, with the exceptions of Florida and Tennessee (both established state highway departments in 1915), South Carolina (1917), Indiana (1919), and Mississippi (1922). And Texas.

(TxDOT Photo)

The age of the automobile arrived in Austin in 1905 with the paving of Congress Avenue, the city's main thoroughfare. (TxDOT Photo)

Even most of the rural South had more advanced road management systems than Texas. Given that Texas also possessed an area that covered 265,896 square miles, a population of over 3 million residents, and less than 3,000 miles of "improved" roads, this fact should have shaken its citizens into action. The reasons for Texas' laggard position were chiefly political. The resistance and controversies stirred by efforts to change the system and legislate for a state-controlled and operated agency arose from the county power structures: across the map, county commissioners and judges fought to retain their rights for road upkeep, the funding generated

Construction of the new Congress Avenue Bridge looking south across the Colorado River in Austin. The new bridge opened in 1910. It was designed by Ernest Emmanuel Howard, the "H" in HNTB. Upon completion, the old iron-truss bridge was removed. The reconstructed bridge was renamed in honor of Governor Ann Richards in 2006. (Courtesy of Austin History Center)

through them, and the decisions on where to spend it, as granted by the cumbersome, government-limiting, and hard-to-amend Texas constitution. Sometimes factions within the counties stood in direct opposition to one another. Cities and towns resented monies spent on other communities that had been generated through bonds and taxes paid for by themselves. Adding to those issues, self-determinism, an inherent stance in the Texas character, flared up like a gas torch whenever county rights seemed threatened.

But universal need outweighed individual interests. The ongoing drama of Texas roads opened in 1903, when the first of a series of attempts toward progress, two road bills, were introduced to the state Senate: one to refine the role of county convict labor in road construction; the second, more important, proposing the creation of a State Bureau of Public Roads. Governor S. W. T. Lanham, the last Confederate veteran to occupy the governor's seat, addressed the legislature, urging the necessity of road improvement. The second bill, however, sponsored by Senator James J. Faulk of Athens, Henderson County, received little support and died in committee.

The next thrust came from Representative O. P. Bowser, who in 1905 introduced a bill to appoint a state expert engineer from the civil engineering professorship at the Agricultural and Mechanical College of Texas. This person would supervise road development throughout the state, as well as regulate the scope of the county commissioners. When the Committee on Roads, Bridges, and Ferries recommended that it not pass, their parry was successful. In 1907 Representative Clarence Gilmore again drafted such a bill, this time specifying the state engineer's terms of service, powers, duties, and salary and actually declaring a state of emergency. The Committee on Roads, Bridges, and Ferries agreed with its framework. Nonetheless, the bill died on the Speaker's desk—canceled due to underadvertising in a sufficient number of newspapers for a sufficient number of weeks, as defined necessary by the constitution.

In 1909, Governor Campbell proposed laws for a "more efficient road system," but again the legislature refused to authorize a department. Representative John T. Briscoe proposed a bill "to provide for the appointment of a Commissioner of Highways." The Speaker of the House now joined the Committee on Roads, Bridges, and Highways to reject it. With both parties united, it was murdered by amendment.

On every side, desperation goaded resolve. The sixth road reform attempt occurred in 1911, with the most sweeping bill yet presented before the Senate, asking for a state highway engineer, a state highway department, and automobile registration fees. Its grave was dug in committee also, along with another road bill that year—the seventh effort.

In 1913, Governor Oscar Colquitt recommended legislative action to address the highway problem but then vetoed a sponsored bill already

Repairing new asphalt in Houston, 1911. (James E. Pirie/TxDOT)

Boy Scouts from Austin aiding in work on Good Roads Day in Travis County, November 14, 1914, when Governor Colquitt plowed and shoveled with Austin businessmen to improve a public highway. (Courtesy of Texas State Library and Archives Commission)

passed by the legislature establishing both a highway commission and a state engineer. Another 1913 bill also died in committee. Two more bills promoting the creation of a highway department—one of them soundly endorsed by the Texas Good Roads Association with the rally cry "Get the farmer out of the mud!"—fell victim, the first to committee and the second to the march of time, when the session calendar ended before passage could be concluded.

Eleven efforts in eleven years: it seemed possible that the stubborn and exacting Texas character might prevail against change forever. And then the balance of interest shifted once again. This time a new enticement arrived to weight the scale: outside money.

The Federal Government to the Rescue

The year 1915 saw several changes in the state of Texas that would prove pivotal over the next two decades. The federal government sent General John J. Pershing to wage war against Mexican Pancho Villa and his revolutionary guerrilla forces following Basilio Ramos Jr.'s arrest and the discovery of the Plan of San Diego. The subsequent massive increase of newly inducted Texas Ranger companies permitted untrained and ill-intentioned men to bear arms with authority and conduct atrocities along the border. A disastrous hurricane struck Galveston and Houston, testing the new seawall in Galveston and altogether killing at least 275 Texans. But equal among these events, if not primary, was the ascension to the governor's chair of James Edward Ferguson, who had been elected the previous November. Thus commenced an era of public infamy unsurpassed by any in the history of Texas leadership since—and road construction played an essential role in its action.

For many reasons, not the least of which were military in concern, the government in Washington had now been galvanizing all the states that had not done so thus far to create highway departments. Foremost in this initiative was Senator John Hollis Bankhead of Alabama, who, along with Representative Dorsey Shackleford of Missouri, in 1914 drafted and presented a bill that designated $75 million in federal funds, apportioned according to a trifold equation, to each state willing to match the funds from its own coffers, the total of which would be earmarked exclusively for the development of highways. The bill provided for the funding of up

Governor O. B. Colquitt is at the plow, and UT president S. E. Mezes is the tall man in white shirt and black tie handling a shovel. The dignitaries were on hand to observe Good Roads Day in November 1914. (Courtesy of Texas State Library and Archives Commission)

to 6 percent of roads statewide over a five-year period. The American Association of State Highway Officials (AASHO) provided the model template for the bill. Stipulations in the bill included that the federal funds would not be released until state legislatures concurred with these provisions: all federal funds had to be matched dollar for dollar through an established, state-run highway department, which would in turn work with the Federal Bureau of Public Roads in reviewing each county's construction designs, road locations, and cost estimates. These last three points were considered paramount, as they meant that roads would then be built with some uniformity and that the roads of adjoining states—in Texas' case, New Mexico, Oklahoma, Arkansas, and Louisiana—would connect at their boundaries, Also, the presumption included that through this mutual cooperation, with its resultant free exchange of information, costs would be minimized and better road-building techniques would ensue.

President Woodrow Wilson, a devoted road advocate, signed the Federal Aid Road Act, also known as the Bankhead-Shackleford Act, into law in July 1916. That same year, route design and construction on the Bankhead Highway began—a project that spanned the entire country, state by state, and ultimately linked Washington, D.C., in an unbroken line with San Diego, California. Wilson's 1916 party platform included this declaration: "The happiness, comfort, and prosperity of rural life, and the development of the city, are alike conserved by the construction of public highways. We therefore favor national aid in the construction of post roads and roads for military purposes."

The equation on which the appropriations would be divided among the states was based on three considerations:

One-third in the ratio which the area of each State bears to the total area of all the States; one-third in the ratio which the population of each State bears to the total population of all the States, as shown by the latest available Federal census; one-third in the ratio which the mileage of rural delivery routes and star routes bears to the total mileage of rural delivery routes in all the States at the close of the next preceding fiscal year.

Once someone studied a continental map, even the most belligerent disdainer of federal or state oversight and input could do the arithmetic and clearly see what this meant: by far the greatest beneficiary would be Texas.

(TxDOT Photo)

Out of the Mud and into the Mire

The Ferguson Years

An earth road is a very desirable type of road so long as it is dry. . . . Its disadvantages are principally due to the effect of wet weather, when it is practically impassable.—R. G. Tyler, "Roads and Pavements: Earth Roads," University of Texas Bulletin No.1922, April 15, 1919

An East Texas road can be defined as "a ditch with a fence on each side, along the bottom of which we have to travel." —George A. Duren, Texas state highway engineer, quoting Jerry Debenport, Dallas Morning News, October 14, 1917

After the Federal Aid Road Act passed successfully into law, the Good Roads campaigners throughout Texas stepped up the pressure to establish an official state highway department. They held mass meetings in multiple counties and vigorously lobbied the government. These efforts also included a visit from the Legislative Committee of the Texas Good Roads Association to Governor James Edward Ferguson (known by his own preference as "Jim"), who had taken office for the second time in January 1917.

Ferguson did not like the Good Roads Movement. In fact, he disliked any person or organization from which he could not squeeze money. Roads, traditionally a county rather than a state responsibility, could not (yet) link directly to his bank account. He never learned to drive an automobile during his lifetime, scorned the notion of any vehicular speed above 10 miles per hour, and denounced those who wished to move faster as "speed maniacs," peevishly demanding a 10-mile-per-hour speed limit. Although technically a lawyer, he had never attended law school. He had won admission to the bar through an oral examination administered by three of his father's oldest friends, the substance of which consisted entirely of a four-person whiskey toast to the departed Reverend Ferguson's memory. As his father had largely

Jim Ferguson (center) with bank officers in front of the Temple State Bank, circa 1906. (Courtesy of the Bell County Museum. Used by permission)

been absent from the family home on his ministerial circuit rides and had then died when Jim was only four years old, it is possible that his memories of this patriarch were dim or specious at best.

Shortly after his brief and blundering attempts at law practice, Jim decided other enterprises would bear more rewards; he wooed with great determination a step-cousin who would inherit a nice property, accepted the gift of a diamond ring from her wealthy widowed mother as a token of gratitude for his help and business advice (boons that he apparently denied his own hardworking, widowed, and impoverished mother after having abandoned her when he was sixteen), and dabbled first in real estate and then in banking, after the step-cousin said yes. A few years later, he entered politics as a new field of opportunism but with no professed interest in transportation—or anything else much, except as a front with which to snare financial support and votes.

However, Ferguson had been elected to the governorship in part due to his reputation as a friend to farmers (as well as a richly rewarded ally to beer brewers and therefore an anti-Prohibitionist). When the Texas Good Roads Association presented him with the facts and statistics, he had no choice but to admit the advantages farmers would reap from an improved road system. Rather than lose his main constituency, he reluctantly consented to the road advocates' persuasions and promised to recommend the creation of a highway department, a highway commission, and the office of state highway engineer to the legislature.

In January 1917, House Bill 2 incorporating these objectives was introduced to the legislative body and passed in both House and Senate. On April 4, 1917, Governor Ferguson signed the bill into law—the only significant legislation enacted by his administration during his entire tenure in office. Texas had at last capitulated. It was now united, in a neck-and-neck tie with South Carolina, for *number forty-fifth out of all forty-eight states* to establish a road authority and to comply with the federal guidelines. South Carolina's legislative body approved founding their own commission on January 20, 1917—almost simultaneously with Texas' bill, which passed sometime between January 9 and the end of the month.

But ominous clouds were gathering over the new department and over Ferguson himself. Another two important House Committee tasks addressed during that same 35th Texas Legislature Regular Session, which met from January 9 to March 21, 1917, bore these titles: "House Committee on Charges against James E. Ferguson, Governor, Investigate—35th R.S. (1917)"; and "House Committee on Investigation of Impeachment Charges against Governor James E. Ferguson, Committee of the Whole—35th R.S. (1917)."

The Senate, at the same time, established a committee to outline a slightly different assignment: "Senate Committee on Formulate the Rules of Procedure in the Trial upon the Articles of Impeachment of Governor James B. Ferguson, Governor: Prepare and present to the Senate not later than Tuesday, August 28 rules of procedure to govern the trial of James B. Ferguson, Governor—35th R.S. (1917)."

_ _

The new Highway Act detailed the structure of the department: It would administer federal funds to counties for state highway construction and maintenance. It would provide for state motor vehicle registration. Administrative control of the department was vested in a State Highway Commission and a state highway engineer. The commission would be composed of three members appointed by the sitting governor (in this case, Governor Ferguson), with consent and approval of the Senate, to six-year terms. Duties of the commissioners included formulating overall policies and plans for a comprehensive system of state and federal highways under the direct supervision and control of the department, cooperating with the federal government in utilizing funds from Congress for road improvement, holding

Automobiles and horse-drawn wagons frequently shared the road. This congested scene is in Upshur County, 1921. (J. D. Fauntleroy/TxDOT)

An early bridge in Limestone County. (E. S. Warner/TxDOT)

public hearings, and outlining geographical divisions within the department. The post of state highway engineer was a literal one, to be filled only by a qualified, certified professional engineer. This person would administer the policies; supervise all location, design, construction, and maintenance of state highways; and direct the workings of the department in general. Convict labor would be used in road construction. The University of Texas and the Agricultural and Mechanical University would both aid the department by testing and analyzing road materials. There would be no more sole county control over road maintenance and construction; instead, counties would supposedly partner with the state.

Registration of motor vehicles was compulsory, and the fees charged per registration were as follows: $3.00 per motorcycle; 35 cents per horsepower or a minimum of $7.50 (equal to approximately $137.57 today) for non-commercial vehicles; commercial trucks were assessed based on weight. All monies generated through registration were reserved for the matching of Federal Aid Highway Act funds. Although he was already under investigation by the legislature for his corrupt practices, it took no time at all for Jim Ferguson to set his eye toward those funds and contrive ways to swing the new department's structure to his fiscal and power-driven advantage.

Organization and Scandal

One of Governor Ferguson's earliest actions, after the passage of the new Highway Department law, was to suggest that all vehicle registration monies —about $1.2 million per year—be funneled through an account set up for the purpose at the Temple State Bank, of which he was the majority stockholder. Since the registration fees had to be turned in to the State Treasury only on a quarterly basis, they could meanwhile be deposited in other banks, particularly his very own.

He also owed that same bank large sums in the forms of overdrafts on his checking account. In addition, he owed mortgage payments on property for which he had duplicated, tripled, and even quadrupled the mortgage debt at other banks—incidentally requesting that the Temple State Bank not list his original debt, in case his political enemies got hold of the information and used it to try to discredit him. Thus, when the young, newly hired president of the Temple State Bank discovered the facts about other outside mortgages on the identical properties, he also found that the bank could seek no remedy regarding default, as its own lien was not listed as first, second, or even third claimant in the right to recover assets. No matter how often he tried to coax or press Ferguson into making even a small payment on his outstanding commitments, the bank president found himself avoided, lied to, or fobbed off as a nuisance. It now seems probable that Ferguson hoped to buy time by using the highway fund deposits to cover those payments, at least for a while.

Texas had elected Jim Ferguson to what some might regard as its grandest position of trust. As a result, he held the authority to appoint a number of officials on its behalf, including the brand-new board of highway commissioners. So, during the scant four months between signing House Bill 2 into law and facing the accusations regarding his malfeasance in court, Ferguson seized that opportunity with relish.

His Highway Commission appointees were Curtis Hancock (chair), T. R. McLean, and H. C. Odle. Hancock was a Dallas lawyer, Oak Cliff city attorney, assistant county attorney, and former legislator who would later devote much of his life to the improvement of Texas highways. T. R. McLean came from Mount Pleasant. Odle, a rancher from the town of Meridian in Bosque County, whose ranch just happened to be situated near Governor and Miriam Ferguson's own Bell-Bosque Ranch, had no road or transportation background whatsoever, nor any history of public service, nor possibly any real interest in either. George A. Duren of Corsicana was appointed as the department's first state highway engineer. They all convened on June 4, 1917, for the first time. In that summer of 1917, after gathering information at public hearings, the commission divided the state into six geographical field districts and proposed an 8,865-mile highway network. The central office locations provided the names of the districts, which, in numerical

One of the first acts of the Texas Highway Department was to create this map. It became the blueprint for planning for years to come. (Courtesy of Texas State Library and Archives Commission)

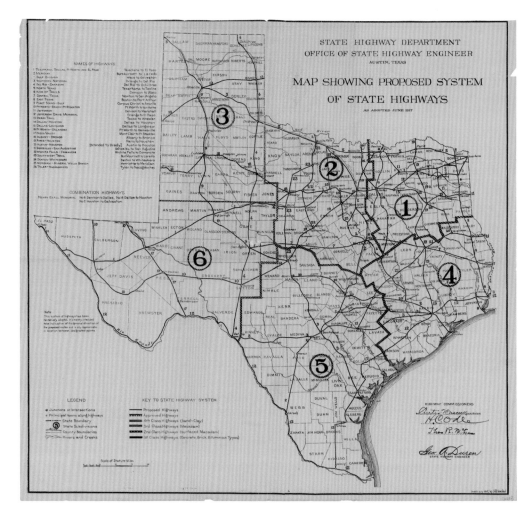

order, were Dallas, Fort Worth, Amarillo, Houston, San Antonio, and San Angelo. Because so many of the men now working on the road programs had formerly been railroad employees who used the railroad idiom "division" rather than "district," that term gradually evolved from an informal usage to a formal title. The first twenty-six highways designated by the department for planning and construction were mapped at this time, based on the responses to the public hearings across the state.

Even before the beginning of that same summer, the truth about Jim Ferguson had already started to emerge. Investigations were yielding cooked books and sworn perversion of facts. Complaints about his shifty financial habits reached the press. The revelations of his con-game property swaps, cash advances on fictitious future crops, mirage business trades and associations, and overdue loans made to him alone by Temple State Bank that actually exceeded the bank's capital now reached the public. Extravagant accounts surfaced, loaded with lists of personal items paid for by the taxpayer—fancy clothing and food, expensive banquets, new decor for the Governor's Mansion, livestock feed for the Ferguson ranch, unauthorized

M arcus Yancey stated in an interview that "the federal government, in 1988, wanted to reestablish the principal arterial system throughout the United States, and each state was required to make its own version [of the map ordered]. When we finished ours, we had about 10,150 centerline miles on the map. And when the Planning Division went back to the George Duren map of 1917, they discovered that he had about 9,700 or 9,650 centerline miles, and you could almost replicate the two [if the second map was superimposed on the first]. His theory was that most of the cities of Texas had been laid out by the railroad system, and it was easiest to follow the railroads to each of those towns. And he did that and then took shortcuts to make sure that you didn't have to go around mountains and so forth. And that map came up and was used for many years as the sole criteria for development."

servants and social secretaries, car repairs, tires, blankets, even a ukulele. Witnesses plunged forward—including the now-incensed bank president. Ferguson's quarrels with the University of Texas boiled over. And for a while, it looked as though his political life and influence would end with his impeachment.

Jim Ferguson was the sole Texas governor ever indicted by a grand jury, until Rick Perry received his two-felony-count indictment in August 2014. Ferguson's indictment by a Travis County Grand Jury in Austin was based on his veto of $1.8 million of legislature-approved appropriations to the University of Texas, when that institution would not dismiss certain regents who had pursued legislative investigations into Ferguson's embezzlement of state funds and improper campaign finance and had refused to fire certain professors for whom Ferguson bore a grudge. The indictment was eventually dismissed, but shortly thereafter, the state legislature went on to vote that Ferguson be impeached and removed from office for a number of additional reasons elaborated in the Articles of Impeachment laid against him or revealed during the trial. Ferguson's particular specialties lay in the embezzlement of public funds to foot his family's personal expenses; the manipulation of multiple mortgages he had acquired by deceiving several different lending institutions, including the one he co-owned; attempted and successful coercion of public officials; lying to the public under oath about his vastly overextended financial debts and commitments; overt cronyism; blackmail; acceptance, diversion, and camouflage of special-interest campaign donations; and a host of other criminal and remorseless actions. One of these elicited testimony that the total amount of funds that had been streamed at Ferguson's explicit instructions through the Temple

State Bank from various state sources and agencies was $760,000 (approximately $13,888,762.50 today).

> Article 8: That James E. Ferguson sought to have the State Highway Commissioner deposit State funds of that department with the Temple State Bank so that said bank might receive the profit and benefit from same, and he being a heavy stockholder, would have received a portion of the benefits. That he also had, or permitted, other departments of the State Government to deposit money with the Temple State Bank . . . said amounts belonging to the State of Texas, and that the Temple State Bank profited from the use of said funds, and the said James E. Ferguson received more than one-fourth of the profit and benefit. (*Journal of the House of Representatives of the State of Texas*, vol. 35, pt. 2, p. 80: The Impeachment Articles of Governor James Edward Ferguson, 1917)

Luckily, Ferguson made the suggestion for highway money diversion to the wrong man, Commission Chair Curtis Hancock, who pointed out to him that Texas already had a perfectly good Treasury designed to shelter exactly those kinds of state funds in an Austin bank previously chosen on the advice of the secretary of state. Although Hancock later testified that he told the other two commissioners of Ferguson's suggestion, he also said that there was never any question of following it. The legislature proceeded against Ferguson in late August 1917, finalizing their decision to impeach with a majority vote on September 26, 1917. One day before, Ferguson hastily tendered his resignation, in (as a preeminent Ferguson historian puts it) an attempt to dodge the inevitable.

According to both public records and private interviews with surviving family descendants, the immediate members of Jim Ferguson's birth family, with the exception of a single sibling, found him so dishonest, treacherous, irresponsible, and manipulative that, after he was expelled from school and had abandoned his impoverished widowed mother, disappearing without a word at the age of sixteen for two entire years, they remained estranged from him for the rest of his life. His childless sister disowned him from inheriting any but the tiniest token fraction of her $300,000 fortune and apparently inserted that $100 pittance merely to preclude or mitigate a will contest—a contest that Jim Ferguson certainly would and did pursue anyway, tying up the estate in fruitless litigation for six years. His own brother hated him and in 1936 carried a six-shooter to the State Centennial celebrations in Dallas, explaining to his young son that it was just in case they ran into Jim.

But unfortunately for Texas, unlike his family, the state had not seen the last of James Edward Ferguson. Nor had it seen the last of the machinations by which he would continue to highjack authority and state contracts; to wrest highway funds, bribes, blackmail, pardon-purchase cash, and monies bullied from highway construction contractors, into his own ever-leaking coffers.

The End of the Beginning, the Beginning of More

Lieutenant Governor William Pettus Hobby immediately replaced the impeached Ferguson in the Texas governor's chair and was elected governor in his own right in 1918, remaining until January 18, 1921. As business writer for the *Houston Post* and the manager and owner of the *Beaumont Enterprise* newspaper, he possessed both commercial and journalistic experience. He was also the youngest Texas governor ever to hold the office up to that point. Meanwhile, the new highway department began to do its job.

George A. Duren, the state engineer, had previously been the city engineer for Corsicana in Navarro County. His employment there marked a return to his hometown; he had been born on family property at Petty's

Paving operation in Cameron County, 1921. (TxDOT Photo)

A narrow bridge between Burleson and Fort Worth on State Highway 2 (called Pat Neff Highway) was the second route designated by the fledging Texas Highway Department but the first one actually completed. (TxDOT Photo)

Chapel, one mile from the Corsicana city limits and very near the site of the first oil well ever drilled west of the Mississippi River—an accidental discovery made in 1894 when the driller was trying to find water. This well launched the first oil boom in Texas, which exactly coincided with the invention of the internal-combustion engine. Until oil was discovered at Spindletop in 1901, the Corsicana fields dominated Texas oil production and prompted the founding of the Magnolia Petroleum Company by Joseph Cullinan. Cullinan then went on to found Texaco in Beaumont and Houston, merging it with Magnolia. The fluke of Duren's birthplace's proximity to the discovery that would fuel the transportation future of Texas now seems a prophetic coincidence.

But the intimate connection to James Ferguson's interests, in the person of Duren's wife's brother, attorney Bryan Yancey Cummings, does not. Cummings was one of several lawyers who defended Ferguson during his impeachment proceedings. George Duren's brother-in-law—a friend, admirer, and supporter of James Ferguson long before his first election to office—had first heard Ferguson speak in Bell County and watched him hold a crowd in thrall for an hour in the rain. There and then Cummings had decided that Ferguson was a man with a future. And a man who, no doubt, might also one day offer good futures to others.

After graduating from the Civil Engineering Department of the University of Texas, George Duren worked for the San Antonio and Aransas Pass Railway Co. and then moved on to the construction of the Dallas-Sherman Interurban line. The oil boom with its flush of new prosperity and opportunities drew him back to Corsicana; the Good Roads Association there, an active and vital group, hired him to help create and maintain several highways into town. It was in all likelihood this experience that contributed to his eligibility for the state job. But Ferguson was never a man to waste a lucrative and important appointment on any person based solely on merit, without the expectation of some kind of return favor. Unlike many of Jim Ferguson's later associates, however, there seems to be no indication that Duren was in league with the governor to harvest more disreputable proceeds from his new position; his public announcement, printed in the newspapers of the time, insisted on adherence to his own policy for awarding contract work strictly to the lowest reliable bidder. This policy not only described the first structuring of what would become the clean alliance between private contractors and the Texas Highway Department, but it would also prove directly opposite to Ferguson's later practices, after Miriam Ferguson was elected governor. In fact, Duren's salary as state engineer was even higher than Governor Jim Ferguson's: he earned $5,000 a year, while the governor's salary reached only $4,000. It was the highest government-sponsored salary in the state at the time—quite a coup for a thirty-six-year-old civil engineer born on a farm to a cattle trail hand.

Duren's first duties included supervising the rush to get vehicles registered. Without money, nothing could be done, and registration meant double funds. He organized ten employees who spent their days in an improvised office inside the House Chamber at the State Capitol Building, handing out stickers that proved motorists had paid and recording the results, at the rate of about thirty-five hundred per day. By the end of 1917, nearly two hundred thousand Texas residents had signed up; in three years, the total would be closer to four hundred thousand. Texans were clearly determined to be drivers.

Meanwhile, Duren kept busy, designing and forwarding other tasks. From his well-written articles and enthusiastic announcements in the newspapers of the period, it seems obvious that he was energetic, competent, deeply involved in the work at hand, and well acquainted with the history, challenges, and prospects of Texas roads. His first achievement after registration was an actual highway—the premier official project under the aegis of the new department. The 20-mile stretch of untreated flexible base was begun in October 1918 and constructed between Falfurrias and Encino, in the same location that now makes up a section of the US 281 corridor. A second highway, this one with a hard bituminous surface treatment 2.5 inches thick built in nine months between 1918 and 1919, ran through Hays County: the first paved iteration of the future I-35. Duren's successful meetings with federal highway officials in Washington produced approvals for seventy-three new state projects involving expenditure of $4,196,421. He also strongly advocated for a paved roadway from Denison to Galveston, via Dallas and Houston, and publicly urged for the training and hiring

STATE HIGHWAY DEPARTMENT – OCTOBER – REGULAR PAYROLL

1917

ADMINISTRATION.

D. E. Colp,	Secretary,	250.00
J. F. Wilkinson,	Head Book-keeper,	150.00
Vance Stockton,	Chief Clerk,	125.00
W. F. Woodman,	Asst. Book-keeper,	125.00
R. J. McMurray,	Cashier,	125.00
N. K. Brown,	File Clerk,	125.00
Jennie George,	Head Stenographer,	100.00
Birdie D. Cannon,	Stenographer,	100.00
Ethel Roberdeau,	Do,	100.00
A. W. Risien,	Shipping Clerk,	100.00
E. B. Kelso,	Book-keeper,	100.00
Chas. Barnhart,	Memo. Operator,	87.50
C. B. Williamson,	Asst. Chief Clerk,	75.00
J. T. Fuller,	Asst. Shipping Clk.	75.00
M. E. Elliott,	Clerk,	60.00
T. A. Plumber,	Porter,	40.00
W. V. Tadlock,	Auditor,	100.00
Willis Carson,	Clerk,	75.00

REGISTRATION.

R. W. Watson,	Registration Clk.	125.00
A. S. Johnson,	Rate Clerk,	100.00
Pete Hodges,	Clerk,	60.00
Hugh Phillipus,	Do,	60.00
Helen Bartles,	Supervisor, Typists,	75.00
Mae Hearne,	Typist,	60.00
Mrs. E. B. Tyler,	Do,	60.00
Vernon Elledge,	Do,	22.62
Frank Wilkinson,	Ctf. Stamp,	60.00
J. D. Jones,	Ctf. Clerk,	100.00
Mrs. M. H. Bradfield,	Clerk,	60.00
Betty Logan,	Do,	75.00
Mome Whitlock,	Do,	60.00
Mrs. D. P. Haynie,	Do,	75.00
Eloise Thatcher,	Do,	60.00
Rochelle Lumpkin,	Stenographer,	75.00
Johnie Moore,	Clerk,	40.00
Maleta Glover,	Typist,	75.00
Werdie Care,	Office boy,	30.00
Otis Johnson,	Clerk,	40.00
Lucile Rawlins,	Typist,	60.00
Bessie Bergstrom,	Stenographer,	60.00
Erna Galbreath,	Do,	60.00
Bertha Johnson,	Do,	60.00
Mrs. W. R. Irvin,	Clerk,	6.00
Mrs. R. J. Blansford,	Do,	60.00
W. K. Hanson,	Do,	46.00
Eulalia Townes,	Do,	60.00
J. W. Bracken,	Watchman,	75.00
Madeline Burland,	Clerk,	43.00
Francis Polhemus	Do,	43.00

1858.12

ENGINEERING.

Geo. A. Duren,	Engineer,	416.66
Julian Montgomery,	Office Engineer,	208.33
Botho. G. Schenck,	Draftsman,	100.00
John D. Miller,	Draftsman,	100.00
Hans Von Carlowitz,	Cost Clerk,	75.00
Bessie M. Sheldon,	Stenographer,	100.00
Heppner Blackman,	Draftsman,	100.00
M. C. Welborn,	Division Engineer,	200.00
John P. Merriweather,	Div. Engineer,	200.00
W. M. Fooshe,	Division Engineer,	200.00

First Highway Department payroll, 1917. (TxDOT Photo)

of more women in road construction, maintenance, and industrial work. Woven into a long, thoughtful essay he wrote for the October 14, 1917, edition of the *Dallas Morning News* describing proposed road routes, costs per mile, and inspection trips he had already made across the state was a vivid capsule history of ancient trails and roadbeds all over the world.

Among Duren's new engagements for the department were three people who would have a lasting impact: Bessie Bergstrom, stenographer, who eventually became the first female chief clerk for Vehicle Titles and Registration; George Wickline, bridge engineer; and Julian Montgomery, whom Duren placed in charge of organizing the department's engineering branch and outlining the state's first bridge and culvert standards.

The first public meeting of the highway commission was held at Mineral Wells June 21, 1917, to designate the highway system. Commission Member H.C. Odle holds dog on leash. At his left in striped tie is Commission Chairman Curtis Hancock. At Hancock's shoulder, wearing a black bowtie, is Commission Member T.R. McLean. (TxDOT Photo)

George Wickline, first state bridge
engineer. (TxDOT Photo)

THE NEW

Highway
Engineer

•

• The selection of Julian Montgomery, of Fort Worth, as state highway engineer focused the attention of the highway world in Texas on the man who will assume direction of the largest and most efficient department in the state government, with thousands of employees and twenty-five division headquarters located at strategic points throughout the state. The choice of Mr. Montgomery also brings back to the Highway Department one of that small band of men who were in the original organization of this now extensive unit.

Mr. Montgomery will succeed Gibb Gilchrist, who for more than ten years has been the executive officer of the department. Mr. Gilchrist will become dean of the College of Engineering at Texas A. and M. The appointment of Mr. Montgomery was announced by the Highway Commission, Aug. 4. In announcing his selection, the commission issued a formal announcement:

"The Texas Highway Commission has accepted the resignation of State Highway Engineer Gibb Gilchrist.

"In choosing a new state highway engineer, the commission has exercised its best efforts to select a competent, experienced and trained engineer and executive of character and ability, who stands high in his profession, who is thoroughly acquainted with Texas and her needs and who is in full accord with the sound principles of high-

way construction and maintenance and who has the facility of expediting the tremendous problems involved in that important position.

"Without application from him for the position, the commission has unanimously selected Mr. Montgomery as state highway engineer and he has ac-

cepted the appointment effective October 1st, next.

"Mr. Montgomery is a native Texan of wide experience and comes to the department at a sacrifice in salary and with a knowledge of the importance of this public trust. He has served as county engineer, division engineer and chief office engineer of the department in the past, and for the three preceding years has been state director and chief engineer of the Public Works Administration in Texas.

"It goes without saying that the Texas Highway Commission is proud of the department's record of accomplishments in recent years, and stands strictly for merit in the personnel of its staff. The commission appreciates and will continue, of course, in the service uninterrupted, those loyal, capable, courteous and efficient members of the department who are devoted to their public duties. There will be no disruption of the personnel of the department. This determination and policy on the part of the commission is fully shared and will be followed by the new state highway engineer.

"Mr. Gilchrist leaves the department as its chief engineer with the appreciation and best wishes of the commission and members of the department as he takes up his new duties. He has rendered to the Highway Department and to the people of Texas honest and efficient service and the commission has exercised its best efforts and judgment in selecting a new state highway engineer who will con-

Turn to page 19

JULIAN
MONTGOMERY

The only known image of Julian
Montgomery in the TxDOT
archives. (TxDOT Photo)

In 1919, a tally of the type and makeup of all the roads in Texas revealed these statistics:

Total miles of road: 148,000
Earth roads: 126,000
Sand-clay: 12,818
Gravel: 7,325
Shell: 1,138
Bituminous surfaced: 395
Macadam: 276
Concrete: 43
Wood-block: 2
Brick: 0.33

Ironically, the tiny fraction of brick pavement grew so large under a later Ferguson administration that it featured as an enormity in one of the most corrupt episodes of the Texas Highway Department's first decade of history.

A bitter disappointment for Good Roads supporters during the department's earliest years was the fact that the federal funds, so eagerly anticipated to start significant road projects across the state, were diverted for a full two years toward military use because of the US entry into World War I. Additionally, the fact that the primary responsibility of road creation and maintenance still lay in the hands of the counties meant that counties paid attention only to their own boundaries, proposing new roads, then terminating many of them at the county line or even before the road reached

Federal aid project in Robertson County, 1921. (J. D. Fauntleroy/ TxDOT)

that far. When Texas had created the new department, it also agreed to the provisions set forth by the Federal Aid Road Act: "that the Federal Bureau of Public Roads would work with each state's department in reviewing each county's construction designs, road locations, and cost estimates." That way, they could ensure that roads were built with enough uniformity so that roads in adjoining states would connect at their boundaries. To this end, a federal review concluded that although the state engineer and the department's field engineers were bound to review county plans and inspect the work, both in process and after completion, each county had its own establishment, and that the Texas Highway Department had so far provided no central planning, organization, or economies of scale. This finding would have a heavy impact on Texas in the near future and engender a threat to the very goals the department wished to accomplish. But the agency was still very young. It had to rise above the turmoil of its genesis, reformulate its structure, get new people on board, and persevere.

After Jim Ferguson's impeachment in 1917, the three-man State Highway Commission he had appointed to serve until 1919 resigned and stepped down a full year early, vacating their seats in January 1918. McClean and Odle faded from the public scene. Curtis Hancock, however, was immediately reappointed by William Pettus Hobby to head the commission. One of the complaints about the first commission (and one that persisted through several commissions to come) had been that the commissioners seemed by and large more interested in finding jobs for their friends than in building a system of roads. The exception to this rule was Curtis Hancock. Both he and George Duren appeared dedicated to the job of designing roads and increasing a competent engineering staff for the benefit of the state. The sad fact was that, of the two, Duren was the one destined to become a casualty of the political earthquake.

In May 1919, Duren still sustained his role in the state engineer's office. Yet for reasons that remain unexplained today but were perhaps related to Hobby's newly reinstated governorship, he submitted his resignation as state highway engineer later that same month. Possibly Governor Hobby preferred the young engineer's replacement, Rollin Joe Windrow, whom Hobby had previously appointed to the Board of Water Engineers. Possibly Duren could not survive the taint associated with his previous boss, a speculation suggested by the fact that the only vocal protests to his removal came from his hometown of Corsicana, in the form of telegrams from his friends urging Hobby to retain him. Instead, he left Austin and moved with his family into a rented house in Marshall, where he had accepted a job as "engineer of public roads" for Harrison County, serving under the supervision of the State Highway Department's district engineer—a dramatic downgrade from his previous position, especially since he had initially hired his new overseer.

A section foreman's car, complete with mascot. (B. F. Sullivan/TxDOT)

Two years later Duren moved to Dallas to work for a private engineering and surveying firm. Two years after that, he was named city engineer for University Park, now an inner suburb of Dallas but at that time a freshly incorporated city on the outskirts and the namesake home to Southern Methodist University. Although he continued to occupy this station until around 1934 and to preserve his Corsicana business interests during that time, he left his wife and daughter behind in Dallas a year or two later and moved north. His last known residence was listed in the late 1930s as LaSalle County, Illinois. From there, he vanished and was not heard from again. To this day his fate remains a mystery.

Duren's wife died in 1958. Her *Dallas Morning News* obituary cited her as "widow of the first State Highway Engineer" and described two of her late brothers' professions but refrained from mentioning her own husband's subsequent positions. Both she and, five years later, her middle-aged unmarried daughter, were buried side by side in a Hearne cemetery.

The Champions of the Public Good

The contrast between the governorships of Jim Ferguson and William Hobby looks, in retrospect, as stark as an x-ray of opposing moralities and standards of integrity. Hobby's embrace of his responsibilities reflected his diligent work ethic, while Ferguson's slippery tricks and exploitations betrayed how adroitly he had shirked any serious work, study, or vocation throughout his life. Hobby's highest priority was to further the progress of Texas and assist Texans; Ferguson's was solely to further his own interests. According to the impeachment findings, Jim Ferguson was now banned from ever holding a Texas public office again. The Ferguson schemes for highway loot had been foiled, at least for the moment; no one at that time could have predicted how they might ever be revived.

When Hobby appointed Rollin Joe Windrow to the state engineer post, Windrow served the Highway Department well and with unexpected effect. Although Windrow would remain behind the desk for only two and a half years, he made a canny hire during his tenure that would eventually change both the structure of the agency and the destiny of all the roads in Texas: he gave a former railroad colleague who had just returned from army service in World War I, a young University of Texas graduate named Gibb Gilchrist, the job of district engineer for San Antonio. There were other challenges, though, that Windrow would soon have to face, and they were underscored by a couple of military excursions.

On July 7, 1919, one month after George Duren's departure from the state engineer's office, a truck train consisting of eighty-one motorized US Army vehicles set out from Washington, D.C., to drive 3,251 miles to Oakland, California, and from there to be ferried across the bay to San Francisco. This debut adventure, known as the 1919 Motor Transport Corps Convoy, intended to test road conditions and military mobility across the country for use during wartime. Its participants included twenty-four expeditionary officers, fifteen War Department staff observation officers, and 258 enlisted men. One of the observers, a young lieutenant colonel named Dwight D. Eisenhower, was charged by the War Department with describing the convoy's progress, delays, and successes.

This wooden bridge was just one of many obstacles faced by the military convoy that crossed Texas in 1920. (TxDOT Photo)

There were not many successes. The truck train included light and heavy motor trucks, touring cars, special makes of observation cars, motorcycles, ambulances, trailers, a tractor, and a machine shop unit. The wooden bridges broken and repaired by the convoy along the way numbered eighty-eight (fourteen in Wyoming alone). According to Eisenhower's report, there were 230 "road incidents" (breakdowns, adjustments, accidents, mud-bog extractions, etc.), no paved roadways from Illinois through Nevada, and nine damaged vehicles and twenty-one men injured en route who failed to finish the trip. After the venture reached its destination sixty-two days later, Eisenhower filed a full report and a day-by-day log revealing the inadequate and often harrowing obstacles to be met by such a military effort. The entire experience made a deep impression on him; he never forgot its lessons. Eventually they shaped the policy of his presidential administration regarding road systems throughout the nation.

More pertinent to Texas and to Rollin Joe Windrow, however, was the convoy that started out from the Zero Milestone in Washington, D.C., less than a year later, on June 14, 1920, and crossed from Texarkana, Arkansas, into Texarkana, Texas, on August 7. This was the Bankhead Highway expedition, called the 1920 Motor Transport Convoy, which would attempt to follow the Bankhead route all the way to San Diego, California, and prove it for military purposes. Although smaller than the preceding test in 1919, with only fifty vehicles, thirty-two officers, and 150 enlisted men, the convoy's troubles and disasters kept mounting the farther it progressed, starting with flooding in Tennessee that continued through the Mississippi "gumbo" mud and on throughout the South, and almost culminating in the nearly impassable sands of Arizona. Certainly the convoy's journey was made merry by the American populace, since despite the many hardships, at every stop along the way townspeople greeted the troops with friendship and held parties, dances, and celebrations in their honor. But the hardships swelled. In Texas, throughout the hot month of August, the heavy machinery labored to navigate the countryside. It took four days just to move from Texarkana to Bonham to Dallas, a distance of 85 miles. Not until September 8 did the convoy achieve the city limits of El Paso. All in all, the convoy's delays required a total of 111 days for it to reach its destination on the Pacific Coast. The officers on the expedition unanimously concluded that the maintenance of a national highway system should be the province of the federal government, not those of the counties working with the states to cobble together a haphazard lacing of roads that led nowhere and often trailed off into a dead end.

After these cross-country test runs were completed and their information evaluated, the newly appointed head of the Bureau of Public Roads, Thomas H. McDonald, in 1921 asked the army to provide a list of roads that it considered of "prime importance in the event of war." The result was an

Inspecting a culvert near San Angelo, 1920. (TxDOT Photo)

Construction takes place in Robertson County along the Old San Antonio Road (part of the de los Tejas Camino Real) in 1921. (J. D. Fauntleroy/TxDOT)

enormous 32-foot-long map, drafted by the Geological Survey and the staff of the Public Roads Bureau and presented before Congress in 1922 by General of the Armies John Joseph Pershing—the same "Black Jack" Pershing who in 1916 had chased Pancho Villa along the Texas border and into Mexico and who had then gone on to lead the American Expeditionary Forces in World War I. The map commissioned by McDonald thereafter became known as "Pershing's Map." Its accompanying request proposed that the US government build 78,000 miles of roads, with an emphasis on coastal and Mexican border defense and the technology, critical strategic priorities, and industrial needs for the military of the period, such as coal fields and iron ports for steel production. Civilian economic development was not a factor for consideration. Consequently, the plan bypassed nearly all of the Deep South, as well as the West Texas and Oklahoma oil fields, which the army had not yet recognized would prove crucial to its long-term requirements. These road proposals became the blueprint for the future interstate system.

In 1921 the Federal Aid Road Act was amended to define more precisely the terms under which fiscal aid was given, and it was then that Texas found itself in danger of having its aid withdrawn. The federal money that war had diverted had begun to pour in again, but if steps were not taken, it would evaporate. No longer could a 20-mile stretch of flexible-base highway be hailed as a badge of pride. Nor could anyone boast about a paltry length of bituminous surface treatment, nor flaunt the new stamped sheet-iron license plates that replaced the homemade numbered leather or wood plates of yesteryear as though they symbolized a great leap forward. Because of the lack of control over the counties and the relatively feeble progress made by the department, the federal government actually pronounced Texas ineligible to participate in a federal project to begin determining the location, design, and construction of interstate highways. Texas, in the

In 1922, as the Highway Department was getting on its feet, a young civil engineering student named Dewitt Carlock Greer was attending Texas A&M University, playing trombone, and developing skills that would impact Texas transportation for decades. (TxDOT Photo)

A paved road in McLennan County in the 1920s accommodates both horse-drawn wagon and automobile. (J. D. Fauntleroy/TxDOT)

government's judgment, needed to show some genuine interest in taking charge over the counties and building some decent roads. Scrambling to meet the new compliance standards, Rollin Joe Windrow oversaw a great expansion of the department's duties. The original six districts grew to eight. Even though Governor Hobby seemed far more interested in education issues than in road encouragement, the department slowly began to work toward its goals.

Meanwhile, during the same summer that the 1920 Motor Transport Convoy was slogging its way across the state along the Bankhead Highway route, another adventurer took to the highways with a lofty ambition in mind. Apparently the State Highway Department had managed to steer the counties toward some modicum of road improvement, for despite rough conditions, gubernatorial aspirant Pat M. Neff was able to mount the first-ever campaign tour across Texas in an automobile, covering 6,000 miles and visiting places that had never before been approached by a candidate. Such access to a possible future governor by people who had never even glimpsed one—meeting him in person, shaking hands, sharing his jokes, swapping stories and concerns, listening firsthand to his speeches—

was a revolutionary innovation, proof of how better roads were already altering lives and shifting fortunes in both rural areas and urban communities. No longer limited to railroad stops on the campaign trail, political hopefuls could now count on personal introductions, make useful contacts according to their own schedules, and canvass for votes even in the most obscure isolated settings. In Neff's case, the result of all this car-travel time and thorough state coverage meant Hobby's defeat and his own victory. And Neff was far more deeply committed to road improvement and its economic advantages than his predecessor had been.

No doubt his car tour had also given him a prime chance to inspect the roads webbing Texas landscapes and assess their conditions and possibilities, because Governor Neff immediately threw his energies into the Highway Department. He made a clean sweep and appointed three new commissioners: R. M. Hubbard as chair, alongside D. K. "Dock" Martin and George D. Armistead. Windrow resigned in 1922, to be replaced by Captain J. D. Fauntleroy. By 1923, only 100 miles of paved road had been completed, although several more projects were under construction or had already been finished, among them the Lone Wolf Bridge over the South Concho River, built under the supervision of Gibb Gilchrist and his latest engineering hire, Thomas J. Kelly. Neff ended up serving two full terms, so he had the opportunity to reverse what he apparently saw as the trend of self-serving apathy launched by Ferguson.

The Good Roads Association had disbanded in 1917, having achieved its aim of seeing the creation of a state Highway Department. Now a group of around two hundred road advocates came to the rescue and founded a new lobbying body: the Texas Highway Association. Early in his first term, Governor Neff addressed them in the Senate Chamber of the State Capitol, promising his own strong support of any legislation that would defer the federal government's stoppage of highway funds into Texas. He charged the organization with one chief priority, his words revealing his low opinion of

This 132-foot truss bridge was rebuilt in 1922 after an accident damaged the original turn-of-the-century structure. It is the oldest surviving bridge and the only remaining metal truss bridge in San Angelo crossing the Concho River from Fort Concho to points south and east, particularly to the Lone Wolf buffalo hunters' camp dating to the 1870s. (Karen Threlkeld/TxDOT)

In 2011, TxDOT and the City of San Angelo undertook a project to rehabilitate the bridge that now serves as a pedestrian landmark on a local hike-and-bike trail. (Karen Threlkeld/TxDOT)

(Karen Threlkeld/TxDOT)

Dressed up for an inspection. The man on the left is thought to be J. D. Fauntleroy. (TxDOT Photo)

the efforts that had gone before: to construct a "big road building program for the state, not a sickly, puny one."

One response to the danger of loss of federal funds and to the new Texas Highway Association's urgings was that in 1923 the legislature voted to raise vehicle registration fees. It also imposed a gas tax of one cent per gallon. Both decisions were intended to help pay Texas' side of highway spending. A speed-limit hike—all the way up to 35 miles per hour—accompanied these measures. But the crowning legislative achievement, the law that would make all the difference, was passed during that same session. It provided that the state take over all responsibility for maintenance of the state highways, thereby wrenching it out of county hands. Several counties, represented by their local commissioners and judges, surged up in indignation and pressed legal suits, challenging this law and refusing to turn over their fees to the state. County money had built county roads; therefore, they reasoned, they had the right to own the roads and keep the taxes they collected for themselves. The tax collector of Limestone County even obtained an injunction. But the court of civil appeals in Waco overturned it and ruled in favor of the state—at last stripping away the longtime impediment and clearing the path for the true beginning of an efficient Texas Highway Department.

One other advancement occurred during Governor Neff's tenure that might one day guarantee the future of the department along ethical, innovative, and cost-effective lines: the promotion, championed by Commissioner Dock Martin and Pat Neff himself, of a replacement for the latest outgoing state engineer, Captain Fauntleroy. For six years, the department had developed in relative political peace. For six years, the Ferguson arm had been crippled, unable to reap any benefits from the Texas taxpayer or manipulate any puppets from behind the scenes. Ten months before the November elections of 1924, Neff requested his Highway Commission to appoint a new state engineer. They picked the San Angelo district engineer Gibb Gilchrist. It was a choice that would draw the lines for the upcoming battle between true road advocates and the corrupt plans of the next administration that was now poised to pounce on every advantage the state had to offer, including highway contracts and all the plums that came with them.

A worker sits atop sacks of concrete on a 1921 federal-aid project in Nacogdoches County. (G. C. White/TxDOT)

Maintenance Foreman H. S. Byrd chats with another crew member, circa 1924. (TxDOT Photo)

Parade in Paris, Texas, circa 1924. (James E. Pirie/TxDOT)

y of Bridges

of Texas, one can cross more than fifty-three thousand structures
ne midair. Some arch over bodies of water—every kind of water,
ling creek that dries up by early summer, to swamps, bayous,
nds; to large artificial lakes studded with ski boats and cypress
weeping, salty breadth of ocean in the Gulf of Mexico—some rise
ssway booming with eighteen-wheelers; some loom high aloft
y used by the occasional Jeep or pickup truck; some merely pave
d "bridge class culverts." And, of course, they cross rivers, the
ers and thirty-seven hundred named streams (also referred to as
All the structures, no matter how humble or brief, officially count
nan-made miracles defying or subverting the bonds of gravity,
red to the earth, sections perched between various types of
into the ground, surfaces hovering over nothingness. Texas has
an any other state, a statistic that should surprise no one. Ohio
d, with more than thirty thousand. And of that multitude of bridges
Lone Star State, two-thirds have been built and maintained by the
nt of Transportation.

The Margaret Hun
Bridge spans the T
River connecting th
Woodall Rogers Fr
to Singleton Boulev
in West Dallas. The
bridge was designe
world-famous engi
and bridge designe
Santiago Calatrava
was opened to mu
fanfare in March 2
second Calatrava
will open across I-3
2017. (Kevin Stillm
TxDOT)

This is the last remaining historic suspension bridge still open to traffic. Text from a nearby historical marker: "This area's first Colorado River bridge was at Regency (now a ghost town), on Mills-San Saba County line. Built 1903, it served ranchers and farmers for going to market, but fell in 1924, killing a boy, a horse, and some cattle. Its successor was demolished by a 1936 flood. With 90 per cent of the work done by hand labor, the Regency Suspension Bridge was erected in 1939. It became the pride of the locality, and youths gathered here in the 1940s to picnic, dance, and sing. Bypassed by paved farm roads, it now (1976) survives as one of the last suspension bridges in Texas." (Kevin Vandivier/TxDOT)

Crossing the Devils F
12 miles west of Del
on US 90. (Jack Lew
TxDOT)

Crossing the Colorado River (Lake Austin), the Pennybacker Bridge was named for Percy Pennybacker, who designed bridges for the Texas Highway Department and was a pioneer in the technology of welded structures. (Greg White/TxDOT)

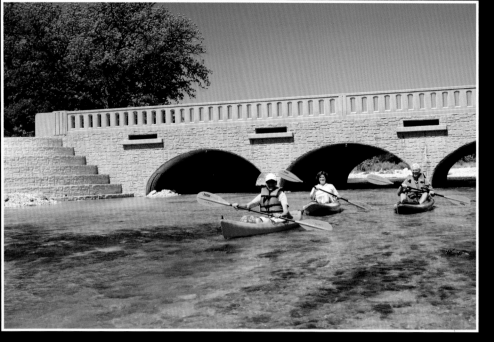

The newly constructed Flatrock Bridge crosses the South Llano River near Junction. The project served to replace an oft-flooded low-water crossing that has been in use as long as humans have inhabited the region. The redesigned bridge improved safety for motorists and provided access to the pristine river for kayakers and other recreational users. (Karen Threlkeld/TxDOT)

Section Foreman Holland Eggleston (standing) and his maintenance crew donned white shirts in 1924 for a formal portrait with their maintenance machines at Bowie. The crew served Clay, Montague, and Cooke Counties with equipment shown here, including Hole and Monarch tractors, two Fordson tractors with maintainers, and a Model T touring car that belonged to Eggleston. (TxDOT Photo)

3 "Fergusonism"

The Ku Klux Klan of 1921 . . . was the most disgraceful outfit that ever lived. . . . We chose . . . the lesser of two evils, although there were plenty of evils during the two years that Mrs. Ferguson served as Governor.—Gibb Gilchrist, "Autobiography of Gibb Gilchrist," 1973

It appeared as if Mr. Ferguson is mad with power.—Speaker Lee Satterwhite, Texas House of Representatives, during corruption investigation of State Highway Department, Dallas Morning News, October 28, 1925

Like James E. Ferguson, Gibb Gilchrist lost his Confederate veteran father at a very young age—three weeks, in his case—and was thereafter raised by a widowed mother. As the youngest of eight siblings, Gilchrist also grew up in straitened circumstances; unlike Ferguson, he developed a strong sense of thrift, a passion for building useful public infrastructure, a loyalty to his family, and an engineer's methodical problem-solving ability. His childhood in Wills Point, in East Texas, also taught him the value of community—a value Ferguson merely viewed as a tool to turn to his own advantage.

After earning his civil engineering degree from the University of Texas in 1909, Gilchrist went to work for the railroad, taking a job as construction engineer for the Gulf, Colorado, and Santa Fe Railway in West Texas. There he gained practice in surveying and route planning, as well as construction repair, enduring the rough conditions and campouts necessary for coping with the terrain. His biggest achievement was to rebuild the Galveston Rail Line between High Island and Port Bolivar after the hurricane of 1915 destroyed it—overseeing the men laboring on the site, bringing in food supplies, organizing the camp, living with sand flies and mosquitoes, dealing with uncertain soil and sandy beaches in the midst of construction—all of which he enjoyed as an adventurous challenge and a source of camaraderie.

When the United States entered the war, he was called to military service as an engineer, during which he rose to the rank of captain. Throughout the conflict he commanded the 540th Engineers, a group of African American soldiers sent to England and France to build roads and rails. The experience left a lasting impression on him, of cheerful friendships, great teamwork, lifelong connections—and music. According to Gilchrist, all of the men of the 540th were apparently musical. A number of them played instruments, including Gilchrist himself. So he formed a band, ordered $1,500 worth of instruments from company funds, and when the order had still not arrived by the time the troops shipped out to return home, instructed that it all be donated to the Paris YMCA. Fifteen years later, the army asked what had happened to the $1,500. Because Gilchrist, as a disciplined, conscientious engineer, still had in his possession all the paperwork involved, he was able to show the authorities exactly where it had gone and why. "There is a moral about this," he said. "Any money dealings you have with the government, you had better keep a good record, because they spend lots and lots of money themselves to see that you do."

This bridge over the Wichita River just could not take the load in 1925. (TxDOT Photo)

Of the company's eventual discharge and farewells at Fort Lee after the Armistice, Gilchrist said with nostalgia and regret: "I saw tears in many eyes, and I think I shed some myself." The army stint also gave him a thorough acquaintance with the types of equipment available to the military for roadwork and construction purposes, knowledge that would come in very useful a few years afterward. Meanwhile, his old friend and railroad coworker from their West Texas days, Hollis Windrow, was happy to engage his services on behalf of the Texas Highway Department.

By the time Governor Pat Neff was guided by Commissioner Dock Martin to officially name him state engineer in 1924, Gilchrist had been working for the department for four years. First he served as district engineer for the San Antonio District and then for the San Angelo District. Like George Duren, he assumed the chief post when he was thirty-six years old. Like Duren, he was energetic, enthusiastic, and determined to make improvements for the state. Unlike Duren, he had neither any connection with, nor tolerance for, ex-Governor James E. Ferguson, a factor that at the time seemed irrelevant.

Now that the counties no longer held the reins for state road improvement and the most basic impediment to the Texas State Highway Department's effective progress had been removed, Gibb Gilchrist faced a second obstacle, just as fundamental as the first. The Highway Department owned not a single piece of construction or maintenance equipment—not even an obsolete split-log drag. Since the counties had always done the work, road paraphernalia had remained their provenance. Most roadwork tasks were still done by hand. Trucks mainly performed as carriers of gravel or asphalt to a job site (although many crews also relied on mule-drawn wagons), but the actual depositing and spreading of these materials was carried out manually, with picks and hoes and shovels, and rakes for smoothing. The work was brutal. In the heat, it was backbreaking. Around some towns and communities, teenagers, both male and female, took summer jobs with the department and, as unskilled temporary labor, accepted the most menial and arduous of the chores. Often the gravel had to be loosened from local creek banks or other sources with pickaxes, dug up, broken down, and shoveled into the wagons, which were built with loose-slatted beds. Once a wagon was full, mules pulled it to the site where the gravel needed to be dumped. Each end of the loose slat boards then had to be wrenched onto its side by hand, under the weight of the broken rock and soil, so that the load would shift into place beneath the wagon bed before getting smoothed with the rakes.

In Johnson City, during the summer of 1924, the labor crew consisted mainly of youths, some as young as fifteen, who were helping complete a 6-mile stretch of road between Johnson City and Austin. One of these, a local boy, had parents who insisted he go to college. For reasons of his own

The 1920s saw the first metal signage replacing painted wooden signs and guideposts blazed on trees. The "L" beneath the highway designations means "turn left at the next corner." (TxDOT Photo)

he rebelled against their desire and instead acted on his father's alternative command to go out and get a job. It was Lyndon Baines Johnson's first official employment. He had worked as a child picking cotton in family's and friend's fields and found that work far too strenuous for his tastes. But in drought-ridden Johnson City that summer, no other choices remained, so he signed on with the state. He, too, strove under the sun alongside his fifteen-year-old friends and classmates, earning $2 a day. He hated the physical toil so passionately that he resolved in future to use his brains instead to make his name in the world. After wrecking his father's car one night, he ran away from home, swearing that he would never do highway work again. He kept his word. The next time Lyndon Johnson made a contribution to the Texas State Highway Department, it would be through directing the use of others' muscle rather than applying his own.

No one knew better than Gibb Gilchrist, supervising all the roadwork in the state, how essential new equipment was. Mule-drawn wagons were all very well, but they would not suffice for the huge amount of work that lay ahead. He needed to supply the department quickly. He designed a plan, commandeering surplus World War I military trucks, tractors, bulldozers, road drags—any useful hardware—from Camp Mabry, the Austin headquarters of the National Guard, and placing an order to lease or buy elsewhere whatever the army failed to supply. Salvaged equipment with few

An early cement truck and steam-powered roller. (James E. Pirie/TxDOT)

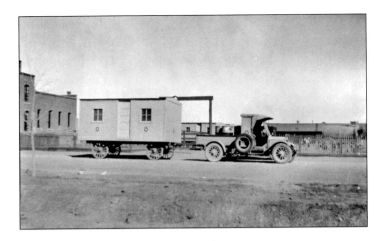

Hooked up to a World War I surplus "Liberty Truck," this portable bunkhouse was home for surveyors and inspectors working construction projects in remote areas. Construction crews camped in tents. (TxDOT Photo)

First state-owned maintenance office in Burleson County, circa 1928. (TxDOT Photo)

moving or workable parts had to be cannibalized to supply other pieces with what they needed.

The department set up an Equipment Division on a corner of Camp Mabry, which led to the birth of the department's Camp Hubbard complex. For this task, Gilchrist had help from the first maintenance engineer, Leo Ehlinger, who also assisted in organizing the department into double the previous number of divisions, bringing the total to sixteen. Before long, thirty-five hundred employees and five thousand contract laborers were at work on road projects throughout the state. But in the midst of all their early progress, disaster struck.

During the years since his 1917 impeachment, Jim Ferguson had been acquiring an income by taking his spectators to court. Always a man to seek an easy living, Ferguson had also found the means of preventing further negative opinion from being published about him, lest the prosecutions he waged get repeated. One by one, he sued newspapers that printed what he claimed to be disparaging stories and comments, each of which he said damaged his reputation. The allegations he complained of did not need to be false but merely unverifiable, in order to earn his vengefulness, justify a legal battle, and produce monetary compensation. He sued the *San Antonio Sun* and received a settlement of $75,000; he sued the *Houston Post* in five separate filings, asking for $100,000 in each case, for its printed observations during his gubernatorial campaign opposing William Hobby. Some of the suits yielded jury awards or settlements (always well below what he asked for); some were dismissed. But the crater of personal debt in which the Ferguson household floundered could not be filled by these awards, so he sought other easy-fix remedies.

On November 8, 1917, less than two months after he tried to flee his impeachment judgment by resigning from office, Jim Ferguson launched the *Ferguson Forum*, a publication that called itself a "Texas newspaper for Texas people" and purported to convey "the Truth Always" to its subscribers—especially farmers and rural dwellers. This four-page weekly could hardly be dubbed an actual newspaper, as it failed to cover current events but instead featured "bait" trade articles such as the one on the front page of the first edition: "Is Cotton a Slacker Crop?". It occasionally touched on farming and livestock problems but chiefly served as a pulpit for Ferguson's opinions, editorials, political bombasts, and attacks against his enemies—including genuine newspapers. Subscriptions to the *Forum* cost one dollar a year. "There are too many gourd heads in the Texas Legislature," Ferguson proclaimed in the first issue. "The bear fight has just begun," he promised in the second issue, announcing that he would run for governor for the 1919–20 term while ignoring the perpetual ban that prevented him from holding any Texas public office. Citing the fact that Travis County had dropped the embezzlement and funds diversion indictments against him because of jurisdiction complications and insufficient evidence to sustain convictions, Ferguson now claimed that all the accusations against him had therefore been groundless. Even after the Travis County district attorney and the judge in the case stated publicly that dismissal did not mean the court held Ferguson not guilty, Ferguson capitalized on the ensuing legal confusion, insisting his was the true version. The Texas Senate was "a kangaroo senate," he wrote, and his impeachment had been illegal.

When he ran against William Hobby in the gubernatorial Democratic primary of 1918 (despite warnings that his run was invalid), he was delighted to win over two hundred thousand votes, although he lost the primary and was removed from the ballot. No scandal could succeed in dampening his political ambitions; no misconduct induced in him a sense of remorse, much less shame. His elaborate financial machinations were now public knowledge, yet he excused them with a number of rationales—for instance, one of the articles brought against him in the Court of Impeachment had involved a "loan" of $156,000 that Ferguson swore on oath bore no political freight but was merely a financial assistance from "friends." When in early August 1918 an investigation by the US Department of Justice proved in court, with sworn testimony from the "donors," that he had received $150,000 plus 5 percent interest in "loans" from three beer brewers in exchange for political favors, the mystery money was finally explained. Ferguson at first denounced the explanation as a lie. Then a few days later he issued a brief public statement, defending the loan as nothing more than what other politicians in similar positions had accepted. Two months later, he transferred the deeds to eight tracts of land that his mother-in-law—the same woman to whom he had given business advice and who had rewarded him with his own diamond ring—had so generously signed over to him and her daughter Miriam in 1907 to the attorney representing the brewers, in cancellation of the promissory notes documenting the loan.

Soon the *Forum* dropped most of its pretense to agricultural advice and evolved into both a polemical organ and a shrewd new source of revenue for Ferguson. He aired his views on women's suffrage (virulently antisuffragist, condemning women voters and their "attendant train of social equality with Negroes, feminism, domination of elections by hypocritical political preachers and union of church and state in an unholy alliance"), called his opponent William Hobby a "sissy," and eventually in 1920 announced his intentions to run for US president; although he could not hold a Texas office, nothing barred him from a federal one. Having divorced himself from the Democratic Party after its state senators impeached him, he denounced it in the *Forum* as "the infernal Democratic Party": "The Texas politicians, all democrats, have not the pride of a jackass or the decency of a skunk." He established his own American Party, but after failing to accrue any substantial presidential support beyond a contingent of elderly rural voters, he dissolved it in 1922. Five days later, he announced his plans to run for the US Senate—on the Democratic ticket. To do so, he had to officially apologize for his denigrations of the party, a gesture so foreign to the conscienceless Ferguson that any semblance of its expression would sound tepid at best. His campaign efforts became effective only when another, more frightening menace than the specter of an impeached governor reared up to take precedence on the political stage.

Meanwhile, advertisements filled most of the space within the *Forum*'s four oversized pages—including ads placed there by Ferguson himself, trolling for investors for his several new oil well ventures (the Money Oil Company, the Chance to Lose Oil Company, and the Kokernot Oil Company), one of which proposed to drill on his Bosque County ranch. Other attempts took place in Liberty and Eastland Counties. He asked his readers to buy shares at $10 each in the Kokernot Oil Company—a name he plagiarized from one of the leading Texas ranching families, no doubt in order to add luster to his enterprise. He hoped to entice as many as six hundred investors by explaining that his lease was only 15 miles southeast of the Humble oil field. "If you can afford it, I want you to buy anywhere from one to ten shares in the Kokernot Oil Company, an association which is capitalized at eighty thousand dollars and is drilling a well on 4,428 acres of land situated 30 miles east of Houston," he wrote. "No honest man can say for sure that he will strike oil. But we are in oil country." Adding that "times are hard," Ferguson warned his newspaper readers to "invest only what you can afford to lose. If we hit the juice a small investment will make you plenty of money. If we don't, just cut out a few pair of them silk sox [sic] and one of them silk shirts, and get some good cotton goods like you used to wear, and you won't know the difference." By 1921, the Kokernot Oil Company had evaporated. If any investors had fallen for his folksy temptations, they must have quickly resumed their cotton socks.

Of more significance, however, was Ferguson's half-ownership of a lignite mine, an asset that he had tried to use as collateral for cash advances on several occasions (claiming that $8 million worth of unmined coal still lay underground within the mine's territory), and the crony ties that it implicated. In September 1919 he sold his share in the mine to Frank W. Denison, the son of his good friend and Temple neighbor Frank L. Denison, whom he had appointed to the State Mining Board while still governor. Not only did the sale of this half-interest produce $35,000 for Ferguson, but an investigation into the appointment of the new owner's father prompted another conflict-of-interest and wrongdoing imputation: it turned out that, while sitting as an officer of the State Mining Board, Frank W. Denison had been awarded the contract to supply the state's lignite needs. This revelation, however, would not by any means signal the end of Frank Denison's link to Fergusonism. In a few short years, his name would become synonymous with fraud in the State Highway Department, once Ferguson at last found the channel through which to highjack department funds and coerce contractors.

The day came when Jim Ferguson's political scheming finally bore fruit. During the early 1920s, the Ku Klux Klan started gaining ascendancy in Texas, its ranks burgeoning so quickly that in some places they soon dominated city and county governments. The Klan attracted civic leaders, judges, politicians, and even members of the clergy into its exclusive, whites-only, Protestant, strictly American-born affiliation, because its harsh Victorian moral code enforcements seemed to counteract the prevailing trends of an increasingly amoral culture. Only a relatively small part of the Klan's defense of morality and society was directed at African Americans. Its campaign of systematic terrorism—beatings and tarrings and featherings—was aimed mostly at bootleggers, gamblers, adulterers, abortionists and abortion seekers, wife beaters, and interracial mingling. In the spring of 1922, the Dallas Klavern of the Klan was credited with having flogged sixty-eight men, most of them at a special Klan whipping meadow along the Trinity River bottom. By mid-1922 it was taking aim at state government, intending to place many Klansmen in elected offices. One of these goals was the US Senate, and another was ultimately the governor's seat.

Since the Ku Klux Klan operated as a cluster of vigilante groups—lynching parties, cross burners, punitive revenge seekers, and supposed social correctors—many outraged people refused to tolerate their actions. Often, though, those actions were never prosecuted due to the judicial power of the members. This new threat to Texan law and order gave Jim Ferguson the target he needed to mount his campaign for the US Senate. Immediately joining the growing wave of anti-Klan protests, he accused anyone who opposed him of Klan sympathies or activities. His negative propaganda even extended to old friends who might give him competition; he seized the opportunity, as usual, to stir up animosity and controversy for his own ends, without consideration for what his opponents actually represented. In the instance of two of his good friends, both running against him for the Democratic Senate nomination, the accusations were true: they were Klan candidates. One of them, Earle Mayfield, won the primary, soundly defeating Ferguson. After his bid failed, Ferguson vowed to see Mayfield vanquished and, in a complete about-face, allied with an anti-Klan group of Independent Democrats and Texas Republicans to endorse one of his longtime enemies in the race. Nonetheless, Mayfield won. It was a victorious moment for Klan partisans. Now the Klan made a more fitting target than ever for Ferguson's political volleys. And he utilized it to achieve his purpose.

After another vain foray into business in January 1923 (he submitted a plan to the legislature, proposing that he lease and manage the state prison system for ten years in exchange for all the state's prison subsidies—perhaps the first-ever attempt at Texas penal privatization) Ferguson resorted once more to the *Ferguson Forum* and its handy podium for promoting his ambitions. Once again he intended to run for governor, once again he solicited donations and investors, this time for campaign funds and *Forum* support. Once again he was foiled by the judicial upholding of his impeachment terms, banishing him from state office. But now he concocted the boldest plot of his audacious career. He instructed his wife, Miriam Amanda Wallace Ferguson (henceforth nicknamed "Ma" by their fans and electoral base, as he would become "Pa"), to file for placement of her name on the ballot in lieu of his. And, eternally obedient to his wishes, she did.

All the antisuffragist and antifeminist sentiments Ferguson had formerly expressed were silenced in this unprecedented move. His talent for quick-change reversals served him well. With no conscience to impede him, he could take whatever position proved most expedient, and the rhetoric he deployed on Miriam's behalf fit beautifully with the picture of a devoted spouse striving to fulfill her husband's ideals. For seven years he had used his family—his wife and two daughters—to wring compassion from the hearts of the public over his pitiful plight as the victim of an illegal frame-up. Miriam equaled reduced taxes, reduced government spending, a self-sustaining prison system, generous fiscal help for rural schools, and salvation for the righteous. And even though Miriam disliked politics, held no aspirations beyond those of housewife, and grew shy and reserved when facing the public eye, she pledged to the citizens of Texas to strike down the Ku Klux Klan.

She also wanted to redeem the family's anguish, which she blamed on the lawmakers who had prosecuted her husband, and to restore the Ferguson honor. "If any wrong has been done, God in heaven knows we have suffered enough," she told her audience. "Though we have lost most of our earthly possessions in these years of trouble, we shall not complain if the people will keep us from losing our family name which we want to leave to our children." She added that, despite her husband having been excessively punished for his alleged malfeasance, "he is conscious of no wrong." She then promised that voters would be getting "two governors for the price of one."

The fluke of Miriam's win during the Democratic primary can be attributed to the fact that, besides Judge Felix D. Robertson of Dallas, the Ku Klux Klan candidate for governor, her two other opponents were both former lieutenant governors in Pat Neff's administration who shared the same last name: Davidson. They were also both anti-Klan. The double-Davidson ticket confused the public and split the vote; the Davidsons—first Lynch, and then Thomas Whitfield—had acted as lieutenant governor consecutively during Pat Neff's two terms in office. Consequently, Miriam edged both of the more popular Davidsons out of the race. Now Jim had a different foe, a menace that superseded any old impeachment articles: Miriam Ferguson would stand as Texas' only chance against Robertson. Meanwhile, the Democrats of Texas had a "rock-and-a-hard-place" choice to make: either nominate Robertson and his KKK agendas or endorse the softer, gentle-toned wife of an impeached governor and public scoundrel. The victor would then run against the Republican candidate. But few cherished any illusions about a Republican's chances. Texas had remained staunchly Democratic since Reconstruction.

It so happened that the state of Wyoming found itself in the throes of a similar question regarding candidates' genders: the wife of its deceased-in-office governor was also running to replace him, but in a specially called election. She, however, was the widow of a beloved statesman, was widely respected herself, and had neither evil forces to combat nor a shameful public record to redeem.

With the battle cry "It is better to follow the sunbonnet than the sheet," Miriam's supporters adapted the song "The Old Gray Bonnet" as her campaign theme tune and summarized the voters' dilemma in the lyric "No masking sheets for me!" The hard choice resulted in a run-off, which Miriam narrowly won. The following November, as predicted, Miriam Ferguson defeated her Republican adversary, the dean of the University of Texas Law School, whom her husband, Jim, had publicly denounced as "a little mutton-headed professor with a Dutch diploma" and whom he claimed was taking orders from the "grand dragon" of the "Realm of Texas," that is, the Ku Klux Klan. Her victory signaled the end of the Klan's dominance in Texas (which had for a short time become the number-one Klan stronghold in the United States), despite the newly elected Barry Miller, a vocal Klansman who would serve as lieutenant governor under both Miriam and her successor. With the Klan defanged, Miller had no bite. "It was all over," a former Klansman reported later. "After Robertson was beaten the prominent men left the Klan. The Klan's standing went with them." In 1925, Miriam Ferguson persuaded the legislature to pass a bill making it unlawful for any secret society to allow its members to be masked or disguised in public—perhaps her sole contribution to the public good, even though the bill's true purpose was the elimination of an adversary for her political gain.

A study in contrasts. One can only speculate as to the emotions of Pat Neff (seated on the right) as he made way for his successors, Miriam and Jim Ferguson.

No state had ever elected a woman governor in a general election before. Texas was now the first to do so. The new female Wyoming governor, sworn into office two weeks before Miriam's inauguration on January 20, 1925, stood solitary in her authority. But no one could mistake Miriam Ferguson for a feminist activist, a combatant for women's equal rights, or a dynamic leader with her own strong vision and agenda; inescapably, the real governor of Texas was Jim.

Only twenty-seven days had passed, when on Monday, February 16, 1925, the newly appointed Ferguson highway commissioners walked straight into State Highway Engineer Gibb Gilchrist's office in the Land Office Building and asked him to act as secretary for their initial meeting. "I'll be glad to," he replied. When the meeting was over, he tendered his resignation. He had held the position for less than a year. "I . . . realized that we had no chance and we had no desire to serve under a Ferguson commission," Gilchrist said. Over the next two years, eight different highway commissioners filled the office. During that same short period, Texas hired and paid the salaries of four different state highway engineers. Gibb Gilchrist accompanied his wife and son to her hometown for a vacation and then joined a private engineering firm and moved the family to Dallas. By the end of Miriam Ferguson's first two-year gubernatorial term, the federal government had cut off all federal road aid and matching highway funds to Texas.

Construction in 1925.
(TxDOT Photo)

The Looting Begins

Two of the new highway commissioners, Frank Lanham and Joe Burkett, were handpicked by Jim Ferguson, who caused his wife to name Lanham, the son of Confederate veteran ex-governor S. W. T. Lanham, commission chair. They partnered with John Bickett Sr., the commissioner Pat Neff had agreed with Jim Ferguson to appoint the previous September to form the governing body to oversee the state engineer and choose highway locations, make planning decisions, and approve contract letting. According to Gibb Gilchrist, who worked with him for three months, Bickett was benign, if ignorant. For this reason Gilchrist addressed his resignation letter only to Lanham and Burkett and significantly left out Bickett's name. "I found Mr. Bickett to be a fine old man of good character, but knowing nothing whatever about highways, how they were handled, or anything of that kind," he said. "Mr. Bickett served . . . throughout the unexpired term of [retired] Mr. Armistead . . . and then he was appointed for a two-year term by Mrs. Ferguson." It soon grew clear that the Ferguson couple and their cronies must have found Bickett's lack of awareness convenient. The number and quality of contracts authorized by Bickett's colleagues would have alerted any person of "good character" who also had the slightest modicum of knowledge about road construction or even a naturally inquiring mind. Although Bickett early on recommended to at least one contractor who had already done sound work for the state that he should continue to apply for road construction work, the resultant bids submitted by the man (every one consistently lower than those of his Ferguson-supporter competitors) were never considered, much less accepted and awarded. When Bickett fell too ill to attend most of the meetings, his absence became invaluable to the co-conspirators.

And thus the march into the designated state highway funds commenced. At this time, Texas had more registered vehicles than any other southern state—over 983,000. It also had higher road expenditures. Because exact procedures for allocation of some of the Texas highway funds had yet to be established, the Fergusons perceived a glorious opportunity to implement their own.

One of Miriam's first acts as governor was to divide the state into three regions, assigning each region as the domain of one individual commissioner who possessed sole administrative control over his region, along with untrammeled power to award contracts within its boundaries, entirely at his own discretion—or so the assumption lay. Rather than function as a unit, the way the legislature originally intended, the commission now worked under the principle of dividing to conquer, without any kind of unanimous decision making or sharing of expertise. As Jim Ferguson was also urging the diversion of highway departmental funds from the state back to the counties, supposedly so that the department's leading responsibility would be to maintain uniformity of construction, the bounty available for each commissioner's fiefdom seemed lavish with possibilities.

It was not until the following autumn of 1925 that many of the transgressions against the Texas Highway Department and the Texas taxpayers began to surface, through inquiries, complaints, and the dogged exertions of one whistle-blower. Yet from the moment Miriam Ferguson took office and the new Ferguson Highway Commission sat down in conference, Jim Ferguson attended their meetings, insisting on closed-door sessions that had previously been open to outsiders and making himself the de facto commander in chief of the official three-person commission that now held absolute control over the $20 million budget for all Texas road projects. As journalist/humorist Don Biggers put it in his book *Our Sacred Monkeys*, "He had no more right to 'sit in' at highway board meetings and boss things around than any other private citizen had." But disentitlement was no obstacle to Ferguson; all real decisions rested within his dictate. The roles of the staunchly supportive commissioners served to reinforce that fact. Two contractors later swore on oath before the House investigating committee that when they approached Highway Commission Chairman Frank Lanham to bid on contracts, he warned them their efforts would prove pointless because the contracts they sought had already been awarded to another firm, the American Road Company, without any notification to the rest of the trade. Lanham also divulged that, although he was indeed chairman, circumstances belied his authoritative position: Jim Ferguson was the person who actually awarded those contracts.

And Jim Ferguson came to the table prepared. Already he had cultivated relationships with a number of contractors by besieging them for ads in the *Ferguson Forum*, some of which cost exorbitant sums in comparison with ad prices in other periodicals and all of which covertly guaranteed future contracts with the Highway Department. The penalty for refusal was no contract work at all. (Ferguson also made it a requirement, once his wife took office, for all state employees to subscribe to the *Forum* at the rate of one dollar per year—not a paltry sum when a pair of girls' good leather shoes cost the same

The Ferguson Forum in its heyday. (Bell County Museum)

amount. Although he disguised this command as a "suggestion," failure to comply could mean dismissal from jobs.) One Bexar County road builder who had been solicited by Ferguson's ad salesman to take a twelve-month, $1,000 ad in the *Forum* and then had canceled the payments on it ten months later after discovering the ad no longer existed, saw his $48,000 maintenance bid turned down in favor of a competitor's bid for $77,000. When asked why he and his partner had agreed to a *Forum* ad in the first place, he explained, "I didn't want it particularly, but we thought it would be better policy to take it than refuse it. We figured it might perhaps hurt future business." He added, "We were afraid we might get in bad if we didn't. . . . By complying . . . we would have an even break with everyone else." Not long after, he and his partner paid $75 for an ad in a special "goodwill edition" of the *Forum*, which he testified they had taken more to "ward off ill will." Road builder F. P. McElrath of Corsicana described how he was finishing several million dollars' worth of state highway contracts when Burkett and Lanham entered their terms on the commission but thereafter was not ever notified when maintenance contracts were to be let—even though he agreed that his firm would be listed in a *Forum* Good Roads advertisement signed by twenty-five or thirty other contractors.

Occasionally the pattern for *Ferguson Forum* ad solicitation changed. Four different contractors were informed a few minutes *after* receiving their awards that they needed to pay between $500 and $1,500 to have their names signed on a Good Roads advertisement in the *Forum*, slated to run for one year. One of them later testified that he was then instructed to go to the Governor's Mansion and see Ouida Nalle—none other than the governor's older daughter—that same day to make his surety bond for the project. Although he replied, when questioned by the House investigating committee attorney, that no, he did not think Ouida Nalle kept an office in the Mansion, he went on to say that he then paid her $650 for a $65,000 bond in the Littlefield Building and that he had been told she was "in the bond business." The ad, he felt, must simply get written off as part of the expense of his contract. "Fergusonism," as it was now called, was plainly a family commerce.

Advertising payments and commissions on surety bond sales amounted to small change, however, in comparison with other prospects for revenue connected to the State Highway Department. Often Jim Ferguson, backed by his wife, directed the commission to award contracts for bids nearly double that of their competitors—apparently for shared kickbacks. At other times the commission assigned contracts for which no bids had been submitted at all. On occasion Ferguson specified particular friends who were to receive contract awards and promised their representatives, including one state senator, that this would happen. In regard to purchases of equipment, including state-owned machinery used for private purposes by nonstate contractors, Ferguson and his commission conceived a plan to outwit the State Board of Control and its refusal to grant certain "emergency" requisitions. Once the board denied approval, the requisitions were withdrawn. Then a rent contract was signed by the Highway Department with various bidders, creating a "rent-to-own" system founded on the mutual agreement that the rented items would belong to the state at the end of a predetermined period. This practice circumvented the Board of Control completely—

The machine shop crew was headquartered at Austin's Camp Mabry, 1926. The men rebuilt and repaired the department's equipment, much of it World War I military surplus. (John Blocker/TxDOT)

The second floor of the Old Land Office Building in Austin (today's Capitol Visitor Center) was headquarters for the chief clerk's office of the Highway Department in 1926. Bessie Bergstrom is pictured on the far right. (TxDOT Photo)

eventually goading its members to volubly criticize the connivers from the witness stand.

In the midst of all the heinous activities, the valid work for the Texas Highway Department was sometimes overlooked. A new bridge that had been erected at large cost and with great difficulty through the cypress swamps surrounding the Neches River near Beaumont was officially opened by Governor Ferguson on May 9, 1925. Work on the bridge had begun early in 1922, as an augmentation of the Old Spanish Trail that, upon the half-mile-long bridge's completion, stretched for 48 concrete miles from the Louisiana border westward toward Nome. Other bridges and roads under construction or near completion at that time were the Brazos River Bridge at Richmond, the concrete road on the Sugarland section west of Houston,

Snow's Ferry toll suspension bridge between Telephone, Texas, and Jackson, Oklahoma, across the Red River was built by Austin Bridge Company in 1927. On December 6, 1940, a suspension cable anchor pulled loose, and the bridge sagged into the water. No one was on the structure when it fell. (TxDOT Photo)

Sowell's Bluff Bridge was built between Bonham, Texas, and Durant, Oklahoma, by the Austin Bridge Company in 1926 and operated as a private toll bridge. The two states paid $60,000 in 1933 and opened the crossing as a free bridge. Eight months later, on February 2, 1934, a steel cable on the Texas side rusted through, and the bridge collapsed into the river. Sowell's Bluff and the Telephone Bridge were two of a succession of toll and free bridges that spanned the Red River, eventually leading to the "Red River Border War." (TxDOT Photo)

paving and bridgework for numerous miles of road throughout East Texas, and the construction commencement of the 3-mile bridge over the Sabine River and its basin between Texas and Louisiana. One year later, in April 1926, another prominent bridge opened, with state governors presiding, respectively, on either end and meeting in Oklahoma for a big celebratory barbecue. This bridge, the first free, untolled span across the Red River on the Lee Highway connecting Vernon, Texas, and Frederick, Oklahoma, became the prototype and precursor of seven others crossing the same river. One of those caused major dissension and "war" between the two states and led to a stand-off manned by the National Guard.

Another highway addition, forged in 1926 when the US government started its policy of numbering national roads with the US designation, was Texas' contribution to the famed Route 66—the grand east-to-west highway that lay north of the Bankhead Highway route and cut across the nation for more than 2,000 continuous miles from Chicago to Los Angeles. In Texas, its path bisected the Panhandle in a neat slice, entering the state at Texola, Oklahoma, and cutting westward through Shamrock and Amarillo to exit just across the border from Glenrio, New Mexico—all of it on preexisting Texas roads through mostly empty countryside.

Some sections of Route 66 remained in the form of simple dirt tracks until the entire route was paved in 1938. Many years later, when I-40 was

US Highway 66 road sign between Jericho and Alanreed in the Panhandle. (Jack Lewis/TxDOT)

constructed, only some sections were incorporated onto the new route, with portions of Route 66 retained as business loops around towns or frontage roads. For now, the Texas stretch of the route was a source of pride for the department: another important national route. By 1927, no fewer than nine sections of major US highways had been established, lacing in a crisscross pattern throughout the state from one border to another to another—both state and international. But the stench of misdeeds and scandal during these Ferguson administration years so overlaid all the good work done by the Highway Department that the department's reputation suffered almost irreparably.

Every situation has a tipping point. After setting so many disparate schemes for profit into motion and sustaining the illusion of their legitimacy, it was inevitable that the Fergusons would find their abuses and lies exposed; sooner or later, the sheer weight of largesse and deception would cause them to topple over. It took eight months into Miriam Ferguson's term of office for this to occur. The crucial day came when the executive secretary of the Texas Highway and Municipal Contractors Association received a request to investigate a private road contractor who was using a piece of state-owned road-paving equipment for a job not under state contract—an accusation that turned out to be true. The secretary, Louis Kemp, visited Jim Ferguson to discuss the problem and request his help in correcting it. Instead, Ferguson gave him a chilly greeting, asking if he was a Ku Klux Klan member or if he supported Miriam Ferguson's administration. When the secretary replied that he owed no loyalties to either institution, Ferguson grew angry and shooed him from the room. Subsequently Kemp went to Attorney General Dan Moody and reported both his concerns about the contractor's wrongdoing and Ferguson's furious reaction. This action prompted the thirty-two-year-old firebrand Moody to take a closer look into state highway affairs. Soon he was opening a Pandora's box of state highway malfeasance and launching his investigation of the Ferguson network's abuses.

Besides the Highway Department, Ferguson had already found ways to milk graft in other arenas. At least one witness, the former mayor of Austin, testified that Ferguson had essentially held an important civic bill hostage. He consented not to have his wife veto it only after he was paid $500. The pardon business also flourished; in her first six months of office, Miriam Ferguson granted a total of 587 pardons to prisoners—36 of them in one day. Her liberality toward convicted criminals provoked wonderment at the time, but later, stories emerged of petitioning friends and family members of the prisoners arriving at the governor's chambers with stacks of bills wrapped in newspapers or tucked inside hats that the wearers doffed only after they stood safely before the desk. The allocation of textbook funds also incited ethical queries, as did the operation of the prison system, which the Fergusons staffed with their own pet appointees and transformed with skimped rations and supplies and suspicious, sometimes cruel, methods of enforcement and routine.

Even Jim Ferguson's nephew, C. E. Ferguson, involved himself in the family enterprises by applying for a salesman's job with a school furniture company, proposing to its agent that they work together in closing contracts for state institutions with the Board of Control. In Temple, on June 27, 1926, C. E. Ferguson confided to the agent that "he had completed arrangements with his Uncle Jim and Roy Tennant, a fellow townsman and member of the Board of Control, whereby he would act as a go-between." The furniture agent later swore in testimony before the House investigating committee that C. E. Ferguson "said there were twenty-six contracts for furniture for State institutions coming up in 1926 and he proposed to let me know when the contracts were to be let. He suggested I come to Austin when he notified me, make the bids, and then pay him the commission." When the agent asked Ferguson if he knew anything about selling furniture, Ferguson replied that he "didn't know anything about the business and didn't want to know." The agent continued: "He [C. E. Ferguson] told me he had made arrangements to split [the money] with James E. Ferguson and Tennant, and he added that his arrangements were 'none of my damn business.'"

But the most glaring cases of fiscal misbehavior were still those involving roadwork. On September 29, 1925, Louis Kemp, the same whistle-blower who had reported the illicit use of state-owned equipment to Dan Moody, spoke before the Association of County Judges and Highway Commissioners in Amarillo of his ongoing research into the State Highway Commission's activities. Previously he had published his discoveries in his own organization's newsletter, but the Texas Highways and Municipal Contractors Association had silenced his commentary after he met with and challenged Frank Lanham's conduct, questioning the cost of many of the state's road contracts and accusing the commission of corruption. Responding to coercion from unidentified sources, the association warned him not to continue his pursuits and fired him from his job when he refused. He persisted at his own expense, convinced now that the web of corruption extended far beyond what anyone realized.

While speaking to the association, he exposed the details of a scheme he had uncovered surpassing the previous ones: a firm called the American Road Company, newly invented in New York on March 27, 1925, and incorporated one day later under Delaware law, had opened an office in Dallas one week after that and in only three days had obtained $2 million surety in Texas road contracts—blossoming into $5 million worth a short while later. These contracts had been secured without bids. The company's sole client was the State of Texas. This was the same company Frank Lanham had mentioned to the two vying contractors eager to bid on upcoming projects when he revealed that Jim Ferguson, not he, had been the person awarding the deals. Later it emerged in court that neither the state engineer nor Commissioner John Bickett Sr. had any part in awarding contracts to the American Road Company, the Hoffman Construction Company, or the other offending companies; both officials had, in essence, been kept in the dark.

Kemp's dogged sleuthing proved very useful to the state attorney general. On further investigation, Dan Moody determined that the American Road Company assets included an asphalt plant purchased from Frank Lanham's own Texas Road Company. According to testimony from F. P. McElrath, the Corsicana contractor who had been squeezed out of state bidding chances, the American Road Company had charged the state 20 cents per square yard to apply asphalt paving—a fairer price for which, he estimated, would have been between 6.75 to 7.25 cents per square yard, based on commensurate techniques and materials. Already, after less than six months' operation, the profits of the company swelled so large that dividends totaling $319,000 had been paid out to company stockholders. When audited, the company's records showed a number of anomalies, one of which was a $3,500 check made payable as cash, which the contractor explained was used to pay promotional expenses. These included a cash contribution to the Ferguson campaign fund.

It grew readily apparent that the Ferguson habit of "campaign fund" bribery acceptance for personal use, whether from brewers or instant entities like the American Road Company, had not changed. Equally important, the revelation that the commission had paid the American Road Company to pave their project with two applications of asphalt, despite their contract's requirement that they apply only one, shone a bright light onto the machinations of both Lanham and Burkett, as well as the Fergusons. When faced with the double-billing evidence to the American Road Company during cross-examination, Commissioner Joe Burkett tried to blame the State Comptroller's Department for not detecting the "bookkeeping" error of a duplicated payment. The comptroller replied that he assumed that when accounts were presented to him, they were "regular." "Vouchers from the Highway Department of the character referred to are shown to have been approved by the division engineer, State Highway Engineer, maintenance engineer, chief clerk of the department, Frank V. Lanham, chairman, and Joe Burkett, member of the Highway Commission, duly verified by affidavit of the payee," Comptroller Sam Houston Terrell stated in defense of his department, thereby indicting the entire chain of authority in the Highway Department, whether intending to or not.

State Engineer R. J. Hank, who had replaced Gibb Gilchrist when he resigned, stated that he had never approved or even known anything whatever of the contracts with the American Road Company and the Hoffman Construction Company until after they were awarded. In conferences with

Burkett and Lanham, he had "recommended that the work be advertised and that the contracts . . . be awarded on the basis of competitive bids. The contracts and specifications for such work were not prepared by me, and at no time were they referred to me for my review or comments." A short time later he resigned his position, to be succeeded by four more state engineers in rapid procession over the next two years—two of them for no longer than a single month. One of those, his immediate successor, found himself abruptly yanked from the job after he disputed a new outbreak of hasty and unnecessary road-sealing contracts, particularly objecting to those being granted to several high-priced companies on a rush-through basis.

For Ferguson's commissioners, and for Jim Ferguson, the exposure of a few dirty deals seemed a good reason to make more—especially if the new ones could then become proof to justify the former. Even after Attorney General Dan Moody sent two letters to the commission chair the following spring, warning him of what would happen if the commission and Jim Ferguson persisted, they still attempted to push the deals forward. "The situation has no precedent in Texas," Moody wrote, "and I do not know whether the people of this State have an adequate remedy in the processes of the law for the prevention of what appears to be an

Dan Moody and his wife, Mildred, 1926. (Courtesy of Texas State Library and Archives Commission)

extravagant expenditure of the Highway funds. If I conclude that there is such a remedy you may rest assured that it will be invoked. In the meanwhile, the responsibility rests upon you and those with whom you have counseled." By December 1926, Moody had found his remedy and made good his threat: he obtained an injunction restraining the Texas Highway Commission from awarding any contracts at all, upon advice and penalty of the federal government as well as the state.

When rumors spread that Ferguson cronies, companies chartered literally overnight by unqualified operators, and *Ferguson Forum* advertisers received almost all the road maintenance contracts, other agencies began to intervene. By now, Kemp, through the articles and exposés he continued to publish, had found champions for his crusade in former governor Oscar Colquitt and former state engineer Gibb Gilchrist, whose expert technical advice and evidence were proving a mainstay in Attorney General Dan Moody's pursuit of criminal activity. Several lawsuits filed by Moody against contracting companies alleged that they accepted payment in taxpayer dollars for work that was never performed at all. By December 1925, the danger signals had reached Washington, D.C. The chief engineer of the Federal Bureau of Public Roads, along with two of his officers, arrived in Austin at the beginning of that month to start auditing the State Highway Department records for misuse of federal road aid funds.

They conferred first with the Fergusons to hear their account and then remained closeted in a two-hour meeting with Dan Moody, who apparently gave them an exhaustive picture of the ongoing prosecutions connected to the American Road Company, including the Fergusons' part in faulty lettings. As this examination continued, the status of Texas' federal aid grew more and more precarious. The government was now on high alert. Further tarnish to the department's conduct was to emerge as the lawsuits progressed. By October 1926, one contractor was testifying before the House investigating committee that Ferguson had demanded from him 10 percent of the bid on a $75,000 road contract and that "he wanted it in $5 and $10 bills." Another witness, an Austin accountant, explained that he had to pay the then-secretary of the highway commission (chairman by the time of the investigation) $350 in cash before he could collect an owed account of $3,620.

In late October 1925, Dan Moody was able to announce that he had recovered a sum of $436,861 on behalf of the State Treasury from a Kansas City bank representing the American Road Company; the company agreed to return the money after Moody had pronounced their dealings with the commission "grossly improvident contracts let at an unconscionable price." On November 20 a further lawsuit ended with cancellation of all of the company's contracts and a total recovery of $600,000. Three more contracting firms soon came under prosecution for shady practices, fraud, contract

Temple Police Chief Sam Hall looks over the "Invisible Track Highway" between Temple and Belton, the largest highway boondoggle in the state's history and legacy of Ferguson-era cronyism. Although it was touted as "the safest highway in Texas," work crews covered and smoothed out the tracks in 1931. (Collection of Weldon G. Cannon and Patricia K. Benoit)

irregularities and discrepancies, and collusion with the commission: the Hoffman Construction Company, the Marine Construction Company, and Sherman & Youmans of Houston. Both Lanham and Burkett were forced to resign from the State Highway Commission as a result of their participation in these and other dubious dealings, some of which paid their holders as much as three times the standard rate for labor and materials. "Probers," as the newspapers called the investigators, unearthed one infraction after another, until soon the piles of evidence were making daily headlines across the state. Slowly the witnesses came forward, often reluctantly, many of them intimidated and fearful of being ruined in retribution.

At the height of the ongoing investigations and their controversy, Miriam Ferguson complained that she was once again battling the Klan. Personal enemies, dishonest contractors, and peevish county officials were also attacking her. Jim took up the banner, defending the same rash of new road-sealing contracts that the state engineer had protested just before he was fired. Jim argued that his wife's administration was simply working diligently with honest firms to replace the inferior roads built by Klan contractors. In his version, the contracting business was so fraught with Klansmen, and non-Klansmen were so rare, that there was no need to advertise for bids since only those few upright citizens could qualify for the jobs.

But when one particular venture drew the spotlight, no amount of Klan accusations could rationalize or excuse its excesses and absurdities. The most egregious case of corruption sanctioned by the Fergusons, as well as the most obvious, was a project called the "Invisible Tread Highway," also known as the "Invisible Track Highway." This highly visible 6-mile length of road connecting Temple to Belton in Bell County (both towns the home ground of the Ferguson family) became the ultimate symbol of all that was wrong with Fergusonism and its effect on the Texas Highway Department.

In late March 1925, the *Dallas Morning News* announced on page 9, in a three-paragraph item, that the "well-known engineer" S. B. Moore of La Porte had just spent the previous Saturday in Austin, conferring with Governor Miriam Ferguson, and that it looked highly possible he would be tendered the position of consulting engineer to the State Highway Department. Moore, it seemed, was an enthusiast of experimental road systems and construction; he was at that time admiring certain "durable road" construction methods he had recently observed in Kentucky. He himself held a patent for another type of experimental road that he believed could prove superbly economical to build and boost safety as well. The design of this road, he admitted, was a little unusual, but it worked on older, tried-and-true physics similar to those of the railroad track, and railroads had certainly withstood the test of time in terms of efficiency and engineering excellence.

The basic construction of Moore's road required a bed of crushed rock overlaid on the road's dirt surface (for instance, the preexisting pike that connected the two towns of Belton and Temple), surmounted by two parallel tracks of 12-inch bricks, spaced apart by the width of a standard wheel base—replicating the parallel tracks on which a train moved forward. The difference was that, unlike steel train wheels with their inner grooves fitting over the steel track rails, the automobile tires that climbed up onto the slope of the "invisible tread" highway must remain firmly and carefully centered on the brick tracks, controlled by the driver without deviation, or else the vehicle would fall off sideways and hit the roadbed. Any vehicle that did not have a standard-width wheel base was ineligible to use it at all. An asphalt covering sheathed the entire construction, presenting a lumpy appearance. But the chances for collision were greatly reduced by this device, as the two-way traffic was accommodated by another set of tracks for cars rolling the other way.

Apparently the Ferguson couple considered it a grand design, one they could endorse—and commission. Because it was experimental, and therefore arguably beyond the prerequisite of the fair bidding process, a host of lucrative opportunities for other kinds of experimentation (in purchasing materials, labor charges, friends' rewards, kickbacks) lay ahead. A short time later, Moore was executing his special highway. The Fergusons' old Temple friend Frank W. Denison, whom Jim had previously appointed to the State Mining Board (and who had during that tenure also enjoyed the contract for supplying the state's lignite needs), now received the contract for construction of the road. As he had no experience in road paving, Denison subcontracted the work to Harold Naylor of the General Construction Company of Fort Worth and sold the materials, equipment, tools, and supplies required by the project from Denison's own hardware store, at inflated prices. These included most of the gravel for the roadbed, which Denison in turn had bought from Jim Ferguson's cousin Fred Ferguson, at a cost of $1.60 per cubic yard instead of the standard or common rate of $1.40 for that grade of gravel.

Another issue was the fact that, although it was new construction, the prodigious expenses, which were more than twice those of the probable original estimate, were charged to the road maintenance fund rather than the road construction fund. When John Ward, the chief bookkeeper for the State Highway Department, expressed reservations about the contractor's invoices that contained a dubious 10 percent upcharge, his superior overruled his protests and invalidated his rejection of certain invoices from Denison and Moore. Ward stood his ground. He still refused to approve payment of the invoices from the highway maintenance account. His superior's response was to pay them without Ward's approval and then to dismiss Ward from state employment—another life damaged or ruined through Jim Ferguson's ruthless financial chicanery.

Meanwhile, the impracticality of the Invisible Track Highway, once opened, quickly became clear: as one eyewitness who rode on it as a car passenger four years later exclaimed, "Once you got on, you had to stay on. You couldn't get off anywhere except at the end. It was the stupidest design you ever saw. It was crazy!" The Fergusons naturally declared it a great success. They traveled by train up to Temple, accompanied by a large party (which included the inept and sickly if well-meaning Commissioner John Bickett, as well as Lanham's and Burkett's replacements, Hal Moseley and John Cage) and formally opened it with much fanfare, a brass band, and a drive down its length and back. In a public speech made in the Temple Theater that afternoon, Jim Ferguson followed Miriam's remarks by announcing that the cost of the new highway, when finally tallied, would be about $26,000 per mile, a savings of $5,000 per mile on other methods of paving. With 19,000 miles of designated highway in Texas, he declared, "if all were paved like this one, the saving to the State will be more than $90,000,000."

The precise original estimates for construction costs of the Invisible Track Highway are now no longer accessible. If Jim Ferguson's declared sums reflected the official estimate (perhaps inflating the initial numbers to cover his back), the total amount of Texas and federal taxpayers' money spent on creating this road nearly doubled it—an average of $49,000 per mile, approximately $3.5 million in today's currency.

After Attorney General Dan Moody's investigations had uncovered as much scandalous conduct as seemed possible, given how many lawsuits now occupied the courts and how long they would take to be resolved, a widespread outcry from various parts of the state demanded that Governor Miriam Ferguson be impeached. Certainly enough negative testimony had now been amassed to deliver a guilty verdict. There could be no doubt of the governor's wrongdoing and her dereliction of her obligations toward the citizens who had elected her. The calm objectivity and solid technical information Gibb Gilchrist had provided, both in the witness stand and behind the scenes, had aided the investigations into the highway affairs. (Gilchrist was later surprised to discover, after all the cases were resolved, that he would be reimbursed a wage of $25 a day for his time and trouble as an expert witness; he had assumed he was merely donating his services and performing his duty as a good citizen fighting corruption.)

The evidence regarding the Board of Pardons, textbook contracts going to the highest bidder, prison abuses, and other secret swindles, was equally cogent. But by the time the investigations were concluded, little remained of the governor's term. Because Dan Moody himself intended to run in the next gubernatorial race and intended if elected to remedy much of the Fergusons' wreckage, he decided against taking action. Two main hurdles stood in his way: the Fergusons still had enough cronies in the legislature to make the process drawn out and highly contentious, its outcome dubious; and he also did not wish such an action to be interpreted as politically motivated. Furthermore, he felt that an ongoing trial would neither benefit the people nor punish Miriam further than the damning publicity had already done. That his view would one day prove shortsighted, with an unfortunate result, he could not have guessed at the time. Of course, Miriam would run again in the upcoming governor's race; of course Jim would insist that she do so, even against her express wishes. Of course, considering the exposures she had just endured, she would lose in the primary—at least this time around. As to redeeming the family name and honor, which had been her originally stated electoral goal: she had now managed to achieve just the opposite.

By February 29, 1936, ten years after that official gala opening, the Invisible Track Highway was in such poor shape that the resident engineer estimated the cost for repair and reconstruction at $38,782.70 (approximately $663,636.17 today) to keep the road in safe and workable condition. He submitted the breakdown of that estimate to the state engineer, who, along with the Highway Commission of the time, made the decision to instead abandon the road to its deteriorative fate, permitting it to remain as a crumbling monument to Fergusonism in its most flagrant manifestations. The Depression now gripped the nation in full thrall. Funds were scarce. The dutiful resident engineer, H. P. Stockton Jr., obeyed his orders. The name of the state engineer who issued them was Gibb Gilchrist.

Two workers employ a creative technique to get a department vehicle out of a sticky spot. (TxDOT Photo)

4

The "Rock Core of Integrity"

I know of no place where the Rangers are needed at this time, unless it should be to form a hollow square around the State Treasury and the Highway Commission.—Gubernatorial candidate Dan Moody, responding to Jim Ferguson's accusation that, if elected, Moody would send the Texas Rangers back to Corpus Christi to enforce the Prohibition Law

Ma Helped Feather Ferguson Nest, Says Committee; Legislature Lays Bare Facts of Texas Administration—Front-page headline, *San Francisco Chronicle*, January 31, 1927

Wish you success and hope the voters . . . will give active support to your campaign. I cannot support Jim Ferguson, proxy candidate. He is my brother, and I know him.—Telegram to Miriam Ferguson's political rival Lynch Davidson from A. M. (Alex) Ferguson

In spite of the bitter campaign battle in the 1926 Democratic primary, during which Miriam Ferguson's supporters attacked the thirty-three-year-old Dan Moody with catchy slogans coined by Jim, like "Dan's the Man, Says the Klan" and "Me for Ma," Moody prevailed over Miriam in the final run-off with a majority of more than 225,000 votes and went on to become the youngest governor ever elected in Texas history.

The Fergusons, with their "compelled advertising" for highway contractors in the *Ferguson Forum*, their pardon practices, textbook graft, and other wholesale incursions on Texas public services, had now become front-page news across the nation, from California through the Midwest to New York, Washington, and New Orleans. This seemed no deterrent for Miriam's busy pen. Throughout her final days in office, she racked up a record number of pardons, even for her—a total of 300 in the two months between Moody's

MIRIAM A. FERGUSON
Candidate for Governor
SECOND TERM
SUBJECT TO THE ACTION OF THE DEMOCRATIC PRIMARY, JULY 24, 1926

1. Read what Fergusonism has done for Texas.

2. A man who will not read both sides of a question is dishonest.—Abraham Lincoln.

3. Do not send a boy to mill.

4. The State is now on a cash basis, and there is money in the Treasury.

5. The penitentiary is paying its way.

6. Taxes have been reduced.

7. No strikes or lynchings.

8. The schools are being run economically and efficiently.

9. The insane have been taken out of jails.

10. Mercy and forgiveness is extended to the friendless and unfortunate.

11. All these a woman Governor has brought to Texas.

12. Why change?

PLATFORM

To the People of Texas:

Some weeks ago in a formal statement I announced for reelection to the office of Governor subject to the Democratic primaries. As I have never scratched the Democratic ticket or violated the party pledge, I felt that I was justified in making this announcement.

In that statement I gave my personal reasons for asking the same second term that has been given to men, and I promised later to give to the people such an account of my administration as would justify my re-election. I now give to the people of Texas that information.

This discussion necessarily involves the statement of past performances and future promises. What has been done and what will be done. Tried by this rule I am willing to be weighed in the balance of public opinion and abide the consequences.

Tax Reduction.

In the first place I promised the people if they would elect me that there would be a reduction in State taxes, and that appropriations would not be allowed to exceed the constitutional limit. To insure the full performance of this promise, I asked the co-operation of the Legislature and stated that I would not approve any special appropriation until the actual needs of the State government were first taken care of and within the constitutional limit. The preceding Thirty-eighth

Legislature appropriated for all purposes $46,855,772.85 and the Thirty-ninth Legislature appropriated for all purposes $36,095,803.95, or a reduction of $10,759,968.90—yes more than ten and three-quarters millions of dollars reduction for the first two years of my administration. To bring about this result it became necessary for me to veto more than two million dollars in appropriations to keep the total within the constitutional limit.

On account of my administration having inherited a two million dollar deficiency (including about $800,000 to pay the debts of the prison system) from the preceding administration, it was not possible to reduce the State rate for 1925, but the constitutional limit was respected and the State rate for 1926 will be reduced from 35 cents to 25 cents—a fraction over 30 per cent reduction.

As a result of these economies there has been brought about a very prosperous condition of the State Treasury. On April 1, 1925, seventy days after I came into office, there was to the credit of the general fund $1,664,372.06, while on April 1, 1926, there was to the credit of the general fund $7,147,904.99, or a gain of nearly five and one-half million of dollars.

It is needless to say that the State can be kept on a cash basis, and there will be no necessity for the discount of warrants drawn against the regular appropriation.

I submit that the promise of tax reduction has been fully redeemed.

(Texas State Archives and Library Commission)

election and his inauguration. The final tally issued during her two-year term came to 3,595 pardons, in a period when the entire prison population of Texas averaged under 4,000 inmates at any one time. According to a *Cleveland Plain Dealer* article printed the day after she stepped down from the governor's seat, the number of Texas prisoners when she first took office had been 3,580. And the total dollars depleted from the Texas Highway

Department's coffer was, in the end, incalculable, compared to what the awarded contracts payments should have been on a fair-market, low-bid basis, if business had been properly conducted. But it amounted to many millions.

And the result? Ten years after the birth and federal alliance of the Texas Highway Department, support for the department came crashing down. The Federal Bureau of Public Roads canceled Texas' road aid funds. It had taken only two years for the Fergusons to demolish the good standing of the largest and richest state agency in the United States. It would take six years of hard toil to clean up the debris, repair the damage, and put the department into its first strong, progressive working condition. And two of those were spent standing up against familiar foes, and keeping them at bay.

Now that the Texas Highway Department's reputation lay in rubble as broken as a dynamited bridge, newly elected governor Dan Moody dedicated himself to rebuilding it. The prime issue on the agenda was to get the federal funding restored. Moody insisted that the system be overhauled and that a "connective and correlated" road plan be developed. To help do this, he immediately appointed a new Highway Commission whom he knew for certain had the best interests of Texas citizens and their road system at heart: Ross S. Sterling as commission chair, allied with Judge W. R. Ely of Abilene and Cone Johnson of Tyler. Ross Sterling was a wealthy Houstonian, the cofounder of Humble Oil Company, proprietor of the *Houston Post Dispatch*, and owner of several downtown Houston buildings. A persuasive and committed road advocate, he felt as outraged and disgusted as Moody

Early roadwork often combined man, mule, and machine. (TxDOT Photo)

It was oilman Ross Sterling's fate to become governor during the Great Depression, the greatest financial crisis ever faced by Texas and the nation. An able leader of the Highway Commission, he put highway construction projects front and center in the plans for creating jobs and boosting the state's recovery. (Courtesy of Texas State Library and Archives Commission)

did. He spent the next several years in first one authoritative position and then another, fighting to correct the destruction wrought by the Ferguson cadre and to stop them from wreaking more.

To launch his reform, he instructed the state engineer to endorse his monthly commissioner's salary check of $267 straight to whichever charity needed it most—a custom he continued even after he lost his fortune during the Wall Street Crash and Depression. For the new Highway Commission's use he furnished a Lincoln and chauffeur and paid for both out of his own pocket to assure his colleagues of fast and easy trips to public meetings and inspection sites, as well as the privacy to confer while they all traveled together. These actions contrasted so sharply with the greed and pillage of the Ferguson administration that there could be no doubts about the intentions of Moody's choices—especially after Sterling also steadily refused to charge any of his commission expenses to the department's accounts.

Ross Sterling's co-commissioners, both lawyers, determined to steer the department back into honest production as well. Moody had thought long and hard about whom to consider for appointment and had even telephoned his old expert witness and consultant Gibb Gilchrist at 1:00 a.m. long distance one night to ask him what kind of highway commissioner he thought R. A. Thompson, one of the most prominent engineers in Texas, would make.

"I urged Dan Moody not to name Thompson or any engineer on the Highway Commission," Gilchrist said—a lucid response to such a late-night call. Moody followed Gilchrist's suggestion, avoided engineers, and picked attorneys Cone Johnson and Judge W. R. Ely to conduct commission business on behalf of the State of Texas. Both were, in Gilchrist's regard, brilliant lawyers and fine, courageous men who would make sure the department would be cleanly run, with no tolerance for bribery, graft, chicanery, or theft. Gilchrist also guided Moody in selecting a new head of the department. "I made a suggestion that he ask the Commission to name [Thompson] State Engineer. . . . The next day, R. A. Thompson appeared in my office in Dallas and proceeded to hit my desk when I had one of these long yellow ruled pads at the point of my pencil as I was getting ready to make some notes. He asked me to consider that I was in his place and was going back as State Engineer, and requested that I put one or more names down for every responsible position. . . . After some considerable thought, I jotted down many names, sometimes alternates, and gave him a full slate."

Moody also felt that the general public should be more thoroughly considered regarding decisions about road priorities. By 1925, more than 975,000 vehicles had been registered through the counties of Texas and their collaboration with the state—a 2,400 percent increase from the previous decade. In 1929, the number would come to more than 1.4 million. The people of Texas must be consulted to set the direction for the future. It was, after all, their money, collected from gas tax revenues and vehicle registration fees, that would help pay for what they needed. So he enlisted the people's opinion and support of his road programs by appointing a thirty-one-member All-State Highway Citizen's Advisory Committee. And through diligence and determination, he and his appointed commissioners convinced the federal government that Texas would comply in every way with every guideline and requirement to meet its standards, thus rescuing the highway aid funds from oblivion.

It was Judge Ely, rather than Dan Moody, who telephoned Gilchrist in the late fall of 1927, shortly after federal aid had been restored and less than a year after Thompson had accepted the position, and asked him to please come back to Austin as state engineer. Rumors had already been floating around that Thompson "had not found the job to his liking" and that Gilchrist might be his replacement candidate. With the Fergusons

Foreman R. E. Wingate and his son Jay show off the first highway signage in Woodville in Crockett County in 1927. The post points the way to Texas 87—two counties away. (TxDOT Photo)

presumably out of the way and no longer a crippling factor, Gilchrist agreed—on the condition that all three commissioners desired him to fill the post. Cone Johnson lay in a Fort Worth hospital, recovering from an illness. The other two commissioners converged on his sickroom with the broad, towering Gilchrist in tow; together the three men loomed over his white-covered hospital bed. Johnson greeted them gladly: here, before him, stood the best cure to all the wounds inflicted by the Fergusons and their limping aftermath. On the spot they agreed to the arrangement. Within a few weeks, Gilchrist had moved his wife and small son from Dallas to Austin. On January 28, 1928, he formally resumed his role as state engineer. "When I went to Austin to relieve Thompson," he said later, "the first thing he did was to reach in his desk and hand me the same sheet of paper I had given him the year before with the remark, 'You should have no fault to find with your organization,' because he had named many, many of the ones I had suggested." That day marked the beginning of a new era.

Evolution of the US 90 High Bridge across the Pecos River

The Pecos River Bridge when it opened in June 1923. It was in service until 1954, when rampaging floodwaters destroyed it. (TxDOT Photo)

The bridge lay in ruin after the flood. (TxDOT Photo)

Construction of the second low-water crossing on US 90 at the Pecos River. The first trestle bridge was washed out by a flood on July 17, 1955. The wreckage of the trestle is in the background. (TxDOT Photo)

Aerial of the second crossing of the Pecos, a temporary bridge built in less than a month in 1955. The wreckage of the original bridge is seen just downriver. (TxDOT Photo)

Construction of the new High Bridge over the Pecos River was completed in 1957. (TxDOT Photo)

Pecos River High Bridge. (TxDOT Photo)

In a 1927 postcard, motorists are seen stopping for a view overlooking El Paso. (Courtesy of John Miller Morris)

Construction continued on state projects; here iron ore is being hauled to a construction site for US 80 in Wood County. (TxDOT Photo)

Glimmers of the new-dawning age encouraged Gilchrist whenever and wherever they appeared. The first time Cone Johnson arrived in Austin after Gilchrist was in place there, he visited the engineer's office and made a declaration of policy. "Gibb," he said, "we are operating the State Highway Department on a cash basis, and if the time ever comes when some voucher is presented to the State Treasury for a just debt of the Highway Commission and funds are not available to pay it, I want you to write a letter to the governor and sign my name and yours, resigning from our respective offices." Gilchrist took this as a powerful lesson and, as he put it, "certainly did watch the purse strings." On many occasions to come, he estimated later, he was holding too much money in the Treasury. "But I didn't want to take a chance of having to prepare that letter." It was, he suggested, his opening glimpse of what would become known as "the rock core of integrity" in the State Highway Department—the cornerstone of the structure they were all uniting together to build.

Legitimate projects continued to get planned. A few that had begun before Miriam Ferguson's inauguration were already under construction or even nearing completion. A new steel-and-concrete bridge opened over the Trinity River in September 1927, connecting Anderson and Freestone Counties—the same river that had presented such peril to the settlers trying to flee Santa Anna during the Runaway Scrape. Multitudes of people from the Trinity watershed listened as Governor Dan Moody made the dedicatory speech and afterward gorged themselves on eight thousand pounds of barbecue. Two months later, Moody helped weld the last link of the Sabine River Memorial Bridge at Orange, Texas, pushing an electric button at the same instant as the governor of Louisiana to join the Old Spanish Trail and the Jefferson Davis Highway on what the press called "the nation's all-Southern route"—and the newspapers said "realized a dream of forty years" by officially connecting the Texas and Louisiana highway systems.

Early map showing the Old San Antonio Road, trans-continental route across the southern United States.

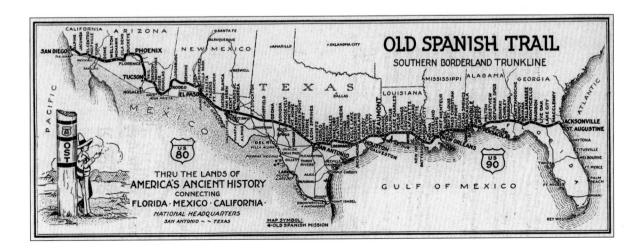

This 1928 photo documents two Denton firsts: the first Model A Ford in town and the first new truck for the county maintenance crew. All previous equipment had been military surplus. Lacking a warehouse, the crew parked their equipment overnight on one corner of the town square. (TxDOT Photo)

In 1926, Texas A&M College had hosted the first Transportation Short Course, a conference in which all department engineering employees could learn about the latest technologies and new knowledge developed by the faculty and researchers in cooperation with the Texas Highway Department, as had been stipulated in the original act. (A&M would not be renamed as University until 1963.) Since then, the Transportation Short Course had grown into what would become an annual event. Now was the time to start putting into action all the new information that Texas A&M and the University of Texas was making available.

One of Gilchrist's primary actions was to once again try to requisition the appropriate machinery and equipment necessary for highway use— $500,000 worth—which, as he and the commissioners pointed out when defending their preferences to the State Board of Control, was $33,000 cheaper than the equipment the board itself wanted to buy. The destinations of all this new equipment lay scattered across the entire territory of Texas. There were 16,000 miles of public highways constructed in the state so far (a figure 3,000 miles under Jim Ferguson's previous declared estimate), but only

Equipment innovation was a key to early highway construction success. This piece of equipment was specially modified to operate in high-water conditions. (TxDOT Photo)

about 1,000 miles had been paved. Of the remaining 15,000 miles, about one-third had either a gravel or a crushed-shell surface. All the rest were dirt. Dan Moody's mission, and that of his new team, was to hard-surface as many roads as possible and to make sure those roads connected smoothly to one another with a uniform finish across county lines.

This meant that the engineer would move to the county seat nearest to the work site, find and cadge some free office space (usually in the county courthouse—often in a basement), round up some furniture to use, or build some himself, and then set up the site of construction before commencing its supervision. Engineering and road crews usually lived in tents, bunkhouses, shacks, or even crude lean-tos in rural areas and empty regions, with no other facilities for many miles. Sometimes the crew would resemble a cattle drive. A cookhouse would be erected to supply hot food, and this entire temporary community housed and serve the engineers and workers shoveling gravel and pouring asphalt. Seventy warehouses were established in different parts of the state so that crews could arrive already provided with the machinery and equipment they needed. Survey teams and engineers often struck out on foot across rocky, rugged terrain that had never seen a road or even a trail, camping out at night under the stars, plotting the route of a new highway that would string one town, such as Llano, to another, such as San Saba, from measurements taken every hundred feet with rods on either side—a return to pioneer explorations. Resident engineers lived in remote Terlingua and Alpine, tending to vast areas of land across sparsely inhabited regions. It was often an isolated and isolating job. And it all had to be paid for.

The predicament of matching federal funds as quickly as possible came before the legislature, which in 1927 voted to increase the gas tax to 3 cents a gallon (lowered to 2 cents in 1928 and raised to 4 cents in 1929). Other solutions also won approval: registration fees were recalculated to charge motorists by vehicle weight rather than horsepower. The average automobile registration now cost $7.87, some 37 cents higher than the original minimum charge of $7.50. In addition, the legislature authorized the establishment of twenty weigh stations across the state that would inspect trucks for weight and safety and charge fees for commercial road use. To supervise these, the department hired twenty plainclothes license and weight inspectors who collected the fines for vehicles weighing more than their registration class allowed. Two years later, in 1929, their number increased to fifty, with men between the ages of twenty-one and forty encouraged to apply to the Highway Department. They would then be redesignated as the Highway Motor Patrol.

By July 1928, Governor Moody was able to announce that all the heavy highway debt incurred by the Ferguson administration, "which could not," he said, "have been paid at ten cents on the dollar with the funds available," was now cleared and that the Texas State Highway Department possessed $11.7 million with which to carry out upcoming work.

Pay as You Go versus Bonds Galore

Throughout the 1920s, the State of Texas implemented a pay-as-you-go policy toward all highway maintenance, road construction contracts, and internal work. Until the depredations of Fergusonism, this policy had presented no problem. Now, even while the state scrambled to replace the depleted funds, that philosophy still held firm; it was further endorsed and cemented in 1929 when, as the entity accountable for all previous county road responsibilities, the state had to pay off over $100 million in road bonds owed by the many counties on the verge of defaulting on their debts.

Dan Moody spent four strong years in the governor's office, during which he solidified sound road plans and procedures and saw much new work accomplished under Gibb Gilchrist's supervision. By July 1928, eighteen months into his administration, 2,400 miles of improved roads had been contracted or were already under construction, more than in any like period in previous Texas history. Laborers throughout the state filled in the gaps between existing roads, and they extended roads that had formerly petered out into dead ends to new destinations. Safety concerns also played a large role in highway design. Engineers frequently encountered much

Maintenance personnel of the Wichita Falls District, posing on the steps of the Wichita County courthouse in 1929. (TxDOT Photo)

The Caterpillar 60 tractor, seen here in Lynn County in 1929, became the department's workhorse for years to come. Two maintenance workers stand next to the Cat on Highway 87 north of Tahoka. (TxDOT Photo)

county opposition when they planned for the 80-foot-wide rights-of-way (soon changed to 100 feet by Moody's commission) that Texas required, in contrast to the much narrower rights-of-way used in other states, such as the 33-foot standard in Pennsylvania. Some roads stretched even wider. In 1929 a four-lane, 150-foot-wide highway was completed between Houston and Galveston. Along many highway sections, signs proclaiming "Slow Men at Work" warned drivers of upcoming construction sites—as well as the fact that the Highway Department had not yet learned how to use effective punctuation. By the fall of 1930, State Highway 2 between Austin and San Antonio, also known as the Post Road, was slated to receive major

Tools of the trade in a warehouse in Graham, 1930. (TxDOT Photo)

Mules remained an essential part of the Highway Department's work as late as the 1960s. This pair, named Frank and Doc, operated a road scraper known as a "drag and fresno" in the Lubbock district in 1929. (TxDOT Photo)

changes, with several death traps and dangerous winding grades in and out of Slaughter Creek and Onion Creek eliminated, new bridges built, other grades evened, and the road straightened.

In the fall of 1929, just a few days after the Wall Street Crash that heralded the Great Depression, the press announced the plans for three new toll-free bridges spanning the Red River between Texas and Oklahoma. Each state would pay half the cost of construction. Overdue projects waited in the list, in every direction. But in October 1930, Ross Sterling, that devoted champion of Texas roads who had helped campaign for 100-foot rights-of-way

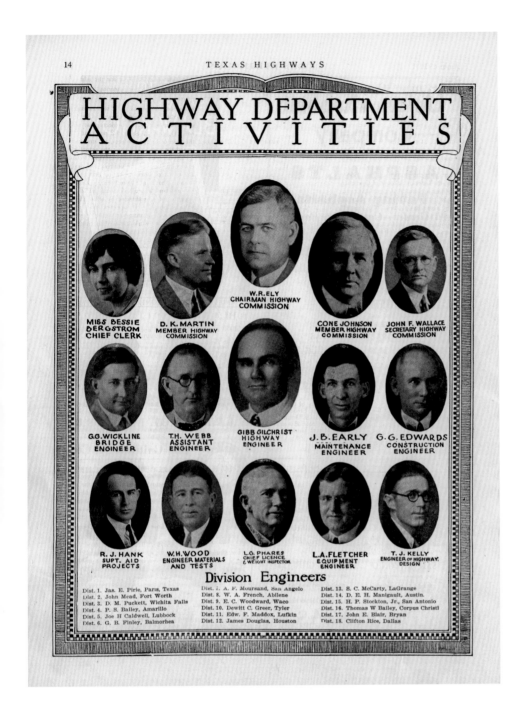

HIGHWAY DEPARTMENT ACTIVITIES

W.R.ELY
CHAIRMAN HIGHWAY
COMMISSION

MISS BESSIE
BERGSTROM
CHIEF CLERK

D. K. MARTIN
MEMBER HIGHWAY
COMMISSION

CONE JOHNSON
MEMBER HIGHWAY
COMMISSION

JOHN F. WALLACE
SECRETARY HIGHWAY
COMMISSION

G.G.WICKLINE
BRIDGE
ENGINEER

T.H. WEBB
ASSISTANT
ENGINEER

GIBB GILCHRIST
HIGHWAY
ENGINEER

J.B.EARLY
MAINTENANCE
ENGINEER

G.G.EDWARDS
CONSTRUCTION
ENGINEER

R. J. HANK
SUPT. AID
PROJECTS

W.H.WOOD
ENGINEER MATERIALS
AND TESTS

L.G. PHARES
CHIEF LICENCE
& WEIGHT INSPECTOR

L.A.FLETCHER
EQUIPMENT
ENGINEER

T. J. KELLY
ENGINEER OF HIGHWAY.
DESIGN

Division Engineers

Dist. 1. Jas. E. Pirie, Paris, Texas	Dist. 7. A. F. Moursand, San Angelo	Dist. 13. S. C. McCarty, LaGrange
Dist. 2. John Mead, Fort Worth	Dist. 8. W. A. French, Abilene	Dist. 14. D. E. H. Manigault, Austin.
Dist. 3. D. M. Puckett, Wichita Falls	Dist. 9. E. C. Woodward, Waco	Dist. 15. H. P. Stockton, Jr., San Antonio
Dist. 4. P. S. Bailey, Amarillo	Dist. 10. Dewitt C. Greer, Tyler	Dist. 16. Thomas W Bailey, Corpus Christi
Dist. 5. Joe H Caldwell, Lubbock	Dist. 11. Edw. F. Maddox, Lufkin	Dist. 17. John E. Blair, Bryan
Dist. 6. G. B. Finley, Balmorhea	Dist. 12. James Douglas, Houston	Dist. 18. Clifton Rice, Dallas

Department leadership team, 1930. (TxDOT Photo)

and upright conduct, stepped down from his position as chair of the Texas State Highway Commission. He had already made his purpose clear to the general citizenry of Texas. He intended to run for governor.

By tossing his name into the ring, Sterling hoped to gain the leverage he needed for a strategy about which he nurtured deep convictions: the financing of highway construction through bonds. The Depression was already placing a drastic burden on the populace. Resistance to bond financing ran high on both sides of the legislature, but without the prop of bonds, Sterling feared the income sources already in place would fail to cover future expenses and the taxation rate would grow punitive. During the first Democratic primary in midsummer of 1930 he stood against ten other candidates—including none other than the discredited ex-governor Miriam Ferguson. That she owned either the will or the energy to run at all could of course be attributed to her function as puppet and proxy for her husband's ambitions. Astonishingly, Miriam succeeded in winning 243,000 votes out of a total 833,000, while Sterling won only 170,000, perhaps thanks in part to Jim Ferguson's sarcastic denunciation of him as a "lordly aristocrat" (appealing for the common man's vote by calling Sterling a rich snob, while ignoring the fact that the Wall Street Crash had now bankrupted him). Sterling's retort would have shamed almost anyone except Jim Ferguson: "I know enough to tell the State's money and my money apart."

Gibb Gilchrist (left) and Governor Ross Sterling (center) inspect construction of the Davis Mountains State Park Highway, more commonly known as the "Scenic Loop." (TxDOT Photo)

In the second primary that August, Sterling defeated Miriam Ferguson by 473,000 votes to her 384,000. At the Democratic Convention in September he was pronounced the gubernatorial nominee; on October 6 he resigned his commission seat, and Dan Moody appointed Dock Martin to replace him, with Judge Ely succeeding him as chairman.

By 1930, Dewitt Greer had joined the Highway Department as the district engineer in Tyler, where he made a name for himself building roads to support the East Texas oil boom. His brown and white Reo was passed down from Highway Commissioner Cone Johnson. Before too long, Greer would be trading his nickname—"The Kid"— for "Mr. Greer." (TxDOT Photo)

Roadside fencing in Calhoun County, 1930. (TxDOT Photo)

After Sterling's inauguration on January 20, 1931, he began lobbying the legislature and Texas voters relentlessly for road bonds. The Depression now slammed Texas hard. Appropriations at this time exceeded revenues, as Sterling had feared, and he had to veto funding for education and other programs. Many good engineers found themselves out of work and, according to Gibb Gilchrist, hungry. Despite the fiscal retrievals under Dan Moody's stewardship, the amount of money to put into actual construction remained limited, but the department provided engineers with other kinds of essential tasks, such as surveying roadways and platting right-of-way maps. Highway crews were expected to supply their own cots and bedrolls at work campsites and to help buy the food, but the department took its responsibility seriously as a resource for Depression-wracked employment. Job creation rather than greatest efficiency became the higher priority, and laborers who were paid between 30 and 45 cents per hour worked no more than thirty hours a week, sharing the jobs in split shifts so more men could be hired to help ease the stifled economy. Hand tools instead of machinery helped make the projects take longer. Mule and horse teams replaced trucks. When in 1931 the legislature cut state salaries by 25 percent, no Highway Department employees complained; too many people all around them were jobless, without any hope of prospects at all. And in the spring of 1932, when the number of divisions (districts) increased to twenty-five and finally stabilized there, the new offices across the state assured more security for more permanent employees than ever before.

Bridge paint truck and crew working on Highway 35 in Polk and San Jacinto Counties in East Texas. (TxDOT Photo)

But there remained in Texas certain citizens who could not at that time benefit from the robust job security and employment practices of the Highway Department. The agency was almost entirely an all-male organization—that is, all white male. African Americans were not hired, even for laborer's work. Neither were married women. The department's philosophy regarding female employees was that, in a time when many households had no earned income arriving at all, it was unfair to supply a single household with two.

Civil Disorder, Border War

Sterling's efforts for bond financing failed. The specter of more debt repelled voters who were aware of how much county bond debt the state was already covering—as an assistance rather than a compulsory assumption of the balance. Meanwhile, big trouble was brewing in East Texas. In September 1930, an oil well christened the Daisy Bradford Number Three in turn christened the tiny hamlet of Kilgore with black petroleum. Another gusher followed it, on a different property. Soon the boom around Tyler, Kilgore, Gladewater, Longview, and elsewhere in smaller Piney Woods communities escalated until the civil peace was shattered. Crime swelled out of control. Thousands of newcomers flooded into the hastily built tar shacks and tent cities outside these towns. While most of the migrants were laborers looking for oil field work, some of them were rowdies and con artists of a different sort: gamblers, bootleggers, pimps, and prostitutes, pouring in to prey on the fresh and teeming populations. To confront the chaos, on August 16, 1931, Sterling declared martial law in four counties for the next six months.

National Guard troops descended on the oil fields to limit waste and regulate the wild insurrection of production that defied the state's conservation laws, while the Texas Rangers swept down to enforce order in the towns. The legendary Ranger Manuel T. "Lonewolf" Gonzaullas, originator of the saying "One Riot, One Ranger," rode into Kilgore alone on his black stallion, Tony, with his pearl-handled, silver-mounted .45 pistols at his sides. Suddenly the downtowns of Kilgore and Gladewater bristled with oil derricks on every building and home lot; a bank was even demolished to make room for a well, and the heads of pumpjacks ducked up and down like feeding animals. Streets lay torn open to accommodate more wells and pipelines. Kilgore's population had increased from seven hundred to ten thousand in two weeks' time. It was now dubbed "the most lawless town in Texas." Ranger Gonzaullas quickly changed that status, shooting and killing three people and acting as sole judge, jury, and jailer for many more. Gonzaullas possessed neither a sense of humor nor patience with criminals, having pledged himself to a life of law enforcement at the age of fifteen after seeing his only two brothers murdered and both his parents wounded by bandits raiding their home. "He'd give you a warning and if you didn't heed it, he'd shoot you. Sometimes he would just shoot for your leg," one oilman recalled.

But Governor Ross Sterling received censure by the federal district court and the US Supreme Court for his "unwarranted" declaration of martial law. The oil boom's challenges created great wealth for leaseholders and drillers, some prosperity for the towns immediately close to the fields, but little help for other areas of Depression-ridden Texas. Cotton prices suffered. Employment dwindled everywhere. The oil industry's trucks and equipment, mushrooming populations, and heavy traffic caused devastating wear and tear on the modest roadways and meager infrastructures of East Texas. This created a new breed of issues (some of which have been repeated recently in other regions of the state, like the Barnett and Eagle Ford Shales with their fracking activities). At a time when the entire nation struggled under hunger and poverty, Sterling grappled with an onslaught of circumstances peculiar to Texas. And in the midst of these problems, the Highway Department contributed another: its very own two-state war.

For several years, the Texas Highway Department and the State of Oklahoma had united in a coalition to build eight toll-free bridges spanning the Red River. One of these, the newly completed, $750,000 concrete-and-steel bridge between Denison and Durant, Oklahoma, was scheduled to open on July 16, 1931. However, starting in 1875, a family named Colbert had operated a toll bridge between the two towns—an interstate bridge that currently carried the traffic for both US Route 69 and US Route 75 across the river. Several other toll bridges and ferries, chiefly family owned and operated, had also offered passage over the river, but their operators had accepted compensation settlements from the two states. The only holdout was the Colbert family, which insisted that the Texas Highway Commission had promised to buy their bridge at the exorbitant price of $60,000 the previous year and to pay the company $10,000 for each month of a specified fourteen-month period in which the free bridge might be opened. Consequently, both states filed an injunction against the Colberts' Red River Bridge Company to prevent it from collecting tolls, in the hope that it would be put out of business. The Colberts retaliated by obtaining a temporary injunction against the Texas State Highway Commission in Houston on July 10, 1931, to stop the new bridge from opening. Ross Sterling's reaction to this legal move was straightforward: he ordered that approaches to the new bridge be barricaded from the Texas end until the legal question could be resolved.

Six days later, the governor of Oklahoma, a colorful character named "Alfalfa Bill" Murray, issued an executive order to open the new bridge. Murray claimed that the land on both sides of the river rightfully belonged to Oklahoma, as defined in the Louisiana Purchase Treaty of 1803. He then

Present-day bridge crossing the
Red River from Oklahoma to
Texas on US 69/75 with the old
railroad truss bridge to the left.
The bridge is located near the
old Colbert toll bridge, which
caused the "Red River Border
War" in 1931. (Stan Williams/
TxDOT)

launched highway crews across the bridge to destroy the barricades, thus committing what the State of Texas regarded as an act of war. In response, Governor Sterling sent three Texas Rangers along with Adjutant General William Warren Sterling, the commander of both the Texas Rangers and the Texas National Guard, to the new bridge to defend the Texas Highway Department workers as they upheld the injunction and rebuilt the barricades that same night. The next morning, Oklahoma crews under orders from Governor Murray attacked and destroyed the northern toll bridge approach from the Oklahoma side, halting all traffic. The people of Denison and Sherman then rose up, rebelled against their isolation, and demanded that the free bridge be opened.

On July 23 the Texas legislature called a special session to pass a bill permitting the Red River Bridge Company to sue the state for the amount claimed in their injunction. The next day, Governor Murray summoned Oklahoma National Guardsmen, declared martial law at the bridge site, and marched through the "war zone" armed with an antique revolver. A few hours later, a Muskogee federal district court issued an injunction forbidding him to block the Oklahoma approach to the toll bridge, thanks to a petition from the Colberts. Murray argued that by holding the position of commander of the Oklahoma National Guard and declaring martial law, he stood above the federal court's jurisdiction. He then instructed the National Guardsmen to allow anyone who so wished to cross the toll bridge. After the Texas legislature passed the lawsuit permission bill, the bridge company joined the state in requesting the court to dissolve the injunction, which it did on July 25. On that day the free bridge was opened to traffic and the Rangers went home.

But on July 27, just two days after the bridge opened, Murray announced that he had learned of an attempt to close the free bridge permanently. He extended the martial-law zone to the Oklahoma boundary marker on the Red River's south bank and stationed Oklahoma guardsmen at both ends of the free bridge. At this action Texans grew outraged, accusing Murray of mounting an invasion and violating their sovereignty. National and international headlines reported the dispute; they presented a picture of civil hostility and instability between fellow countrymen. Finally, on August 6 the Texas injunction was permanently dissolved. The Oklahoma guardsmen were withdrawn. A dedication ceremony involving thousands of participants on both sides of the bridge took place on the morning of September 7, 1931, at which time Governor Murray's daughter Jean broke a bottle full of Red River water on the bridge and pronounced it open "for the free use of all the people." And the penalty for all this upheaval was that Ross Sterling served only a single term in the Texas governor's seat.

The Wolves Gather

During the first Democratic primary in the summer of 1932, Sterling found himself opposed by eight candidates. Jim Ferguson took sly delight in exploiting Sterling's two-year series of ordeals by mocking him with this jaunty little ditty:

> *O I came to the river and I couldn't get across*
> *Because of a row between Bill and Ross.*
> *There stood two bridges, but yet I had to swim.*
> *O why in heck didn't we elect Jim?*

The veil no longer even hung thin; the rhyme made no pretense to hide the true identity of the next hopeful candidate. Its needling arrogance told Texas that once again, in a fourth attempt, Miriam Ferguson's husband intended for her to grind through the campaign to the top. It also implied that Jim Ferguson could have stopped the conflict if he had been elected governor (presumably because he would have had no scruple about ignoring the lawful injunction) and thereby suggested the lack of ethics he was ready to apply in the job. But events particular to the time and their ensuing pressures had created the opportunity. The chance to profit from Sterling's duress must have seemed a heaven-sent gift to the couple. Jim and Miriam scarcely needed to do more than remind the voters of Sterling's dilemmas to place the blame of circumstantial crises squarely on him and to propose themselves as the antidote. When the second Primary Election Day arrived, Miriam Ferguson narrowly defeated Governor Sterling by 477,000 to 473,000, a squeak-by of only 4,000 votes. Because conditions had not changed for Republican candidates in Texas, this meant Ma Ferguson was a foregone conclusion. It also sealed another fateful day for the State Highway Department.

Dock Martin, who had taken Sterling's place on the commission, wrote a letter expressing his dismay to his brother, describing the Fergusons and their friends as "a bunch of wolves." Now that federal funds had been restored, instant action was required by the Highway Commission and the state engineer to protect the hard-won monies in the Treasury already designated for highway use. Gibb Gilchrist and the commissioners rushed through the necessary steps by allocating hundreds of thousands of dollars in federal road money to a host of projects before Miriam Ferguson could be sworn in.

Nonetheless, in late November 1932, as Ross Sterling spent his last weeks in office, Jim Ferguson filed a temporary injunction to freeze those same state highway construction funds, with the plan that the money would remain preserved until the Fergusons could once again grasp it. Although the injunction was later dissolved, Governor-elect Ferguson and her husband went so far as to appeal its nullification. But by the time the courts could process the case, Miriam was back in the Governor's Mansion, and the monies were already in her command.

When Miriam Ferguson took her second inaugural oath in January 1933, Adjutant General W. W. Sterling resigned his office, and an entire group of state employees left with him: forty men in the Texas Rangers. Their action was in protest to the corruption and political patronage that went against the Ranger code of ethics; they had thus openly supported Ross Sterling in the governor's race. Among the defectors was Senior Ranger Captain Frank Hamer, the same lawman who would, as a member of the Texas Highway Department's newly formed Highway Motor Patrol a short time later, put an end to the murdering odyssey of infamous outlaws Bonnie Parker and Clyde Barrow. This in turn led to one of Miriam Ferguson's most notorious actions during the course of her second term. She fired all the remaining Rangers, including even the renowned Lonewolf Gonzaullas, and began appointing her own kind of lawmen, a force that enlisted more than one ex-convict. Other men with shady backgrounds also got to wear the silver star.

Yet the Fergusons were by no means satisfied. They still had more moves to play on the chessboard of power and profit, now that they had finally regained it; and of course they needed to prepare the ground for further scams. The next effort to exert control over the State Highway Department came only one week into Governor Ferguson's new administration. By planting another crony on the Highway Commission, the couple hoped to cancel the effectiveness of the two remaining commissioners still serving out their six-year terms after Cone Johnson's upcoming retirement and maneuver all control into their hands. Governor Ferguson announced her intention to appoint a dear old friend, Frank Denison, as chairman—the same Temple ally who had reaped so many benefits from his previous associations with them, including his old seat on the State Mining Board while enjoying his state lignite contract and, of course, his role as contractor for the Invisible Track Highway between Temple and Belton. The audacity of this move, made while the state legislature was in adjournment, stunned the department. But it depended on Senate confirmation, which was by no means certain. Denison did not care. "In a few days," Gibb Gilchrist said, "Mr. Frank Denison came into my office. He didn't greet me. He didn't shake hands. He just threw a commission [document] down on my desk and said, 'I want to call a meeting of the State Highway Commission for next Monday,' this being about Wednesday or Thursday. I said, 'Mr. Denison, you haven't been confirmed, have you?' He said, 'I have everything they have,' pointing to his commission [document]. I said, 'Mr. Denison, I am going to write the Commission and tell them what you have said, but I don't think they will be down here.' With that he stalked out of the room, and the next day he was rejected by the Senate by one vote—which was worth millions of dollars, really."

Yet Ferguson's audacity did not stop there. It reached toward new heights. The first time Denison's name appeared before the Senate, contrary to Gilchrist's later memories, he was actually rejected by a vote of 19 to 11. But in early February Miriam Ferguson again submitted Denison's name, asking for his approval as Highway Commission chair. This move was unprecedented in Texas; Senate reaction ranged from startlement to fury. In outraged response to her request, Senator Walter Woodward made serious accusations on the Senate floor, repeating rumors that blackmail, coercion, threats, intimidation, and promises had been used to leverage votes for Denison—who, incidentally, was claiming in a letter to the attorney general that he had never actually sought the job and did not really want it unless it was "satisfactory to all parties concerned."

One rumor Woodward quoted was that senators with state-supported schools in their districts had been warned they would lose appropriations if they failed to support Denison; the resubmission, Woodward said, was a test to see if Ferguson's machine had yet suborned enough votes to put

Denison into place. In other words, how strong was their power in general? What else could they achieve through their usual practices of blackmail, pressure, and graft? When asked for an explanation, Miriam Ferguson said she had resubmitted Denison because she felt there had been some confusion and misunderstandings accompanying the first submission.

Meanwhile, to everyone's surprise and chagrin, Denison continued to turn up at commission meetings and sit in, as if he had a perfect right and authority to be there. He persisted in this behavior until Attorney General James Allred barred his attendance for once and for all. But first, the question of his commission still hovered over the Capitol, and now some very annoyed senators were wondering about Miriam Ferguson's inside sources of information. The law clearly stated that the initial vote in executive session be kept strictly confidential. How, then, did Governor Ferguson know of any confusion and misunderstandings that might have influenced Denison's appointment?

The controversy and debate over the validity of resubmission went on and on, in a ruling made by the lieutenant governor and in a closed session during which the Senate once again refused to confirm Denison's appointment. The Fergusons in turn rejected the decision, insisting that only a simple majority was necessary for confirmation rather than the expected and traditional two-thirds majority. Four days later, Denison tried to assert his power as chairman of the commission (the job he claimed he had never sought) by ordering the state comptroller and the state treasurer to dishonor any payroll or other vouchers from the Highway Department that did not bear his signature. Governor Ferguson went further; she asked for the results of the Senate vote in order to support her argument that Denison had indeed received the simple majority she felt should suffice. The Senate turned her down by a vote of 21 to 9, citing a Senate rule that votes taken in executive session were privileged and that therefore she had no right of access. They were upheld in this by Attorney General James Allred. However, not until May 31, 1933, when the Supreme Court of Texas ruled that Denison's so-called commission was invalid because its approval did, in fact, require a two-thirds majority vote from the Senate, did Denison and the Fergusons finally give up the fight.

In June 1933, a new commission member strode into the tall, easy-going Gibb Gilchrist's office. "Then came the news that Mr. John Wood had been named Chairman of the Highway Commission," Gibb Gilchrist recalled. "John Wood came over to see me. He didn't have his commission in his hand because he had already been confirmed. He stuck out his hand, though, and said, 'My, you're a big devil, aren't you.'

"That began a friendship," Gilchrist said. "He was my friend. He was sympathetic to every fair recommendation that I made as long as I was in office. He liked the red-eye [alcohol] a little too much, but he never was under the influence of it where he wasn't big-hearted and fair, and Jim Ferguson didn't run him at all." For reasons of his own, Wood had apparently determined that he would conduct himself with immaculate cleanliness while in office, despite his appointment by, and friendship with, the Fergusons.

Still the Fergusons kept on the attack. The Highway Department was far too juicy a plum to relinquish. They also had a catalogue of vendettas to fulfill. In light of his expert testimony about their schemes a few years before, and his speed in shielding highway funds from their control, State Engineer Gibb Gilchrist apparently ranked high on that list. Even before taking office, Governor Ferguson joined her husband in pointing a finger at the department and accusing it of a $1 million shortage.

The commission's indignant members quickly countered the accusation by presenting specific evidence to the Senate. Behind the scenes, the Fergusons were sending in their henchmen. "A man named Roy Sanderford from Belton, the center of Fergusonism, came to the Senate," Gibb Gilchrist recollected. "He told some of my friends that his only purpose in coming to the Senate was to get rid of me, that he had his knife out and sharpened. . . . I called on Mr. Sanderford after he had introduced a resolution to investigate. I got acquainted with him," which, in Gilchrist's modest language, meant that he disarmed and befriended Sanderson, "but they went on through with the investigation.

"This was a time when the Senate acted as a committee of the whole around their long table in the Senate Chamber. One of the attorneys for the Fergusons, a Mr. Hair of Belton, implied that something Senator Walter Woodward had said was untrue. I was seated right behind Woodward and I saw his neck turn white as a sheet. He reached over and got one of these

Engineers celebrate the arrival of five new Duplex motor graders in the Lubbock District in 1934. The equipment represented a tremendous advance in the department's ability to grade road surfaces and shape road shoulders and drainage ditches. (TxDOT Photo)

heavy cut-glass water pitchers full of ice and water and he threw that right into Hair's face. Blood and water spattered all over me and everybody else. Pandemonium reigned for a little while."

In a turnaround that depended entirely on the impeccable records both the commission and the department had been keeping (and that demonstrated the lesson Gilchrist had learned from the US Army years before during World War I, when he requisitioned band instruments for his unit), the investigation began to swerve in the Highway Department's favor. The accounts presented to the Senate over a three-week period answered for every penny's worth of funds, both present and absent, and the scrupulous care with which they had been managed, amply proving the integrity of the department's personnel—as well as the frivolity of the investigation. It came to an end "after an impassioned speech by George Purl of Dallas," Gilchrist said, "after questioning one of the state auditors who had testified against us, making him admit that the Commission had spent $200 million during the period he investigated; that $196 had been misappropriated or stolen; that the Department had fired the men that stole the money after taking it out of their payroll. The Committee voted then to exonerate the Department and commend it twenty-six to nothing, and my friend, Roy Sanderford, voted with the majority, and soon thereafter began the friendship that has lasted to this day between Roy and myself. He saw that he had been misinformed, and he was man enough to admit it."

Yet another Ferguson chum had now surrendered his agenda to the goodwill and candor of Gibb Gilchrist. In all, Gilchrist's department would be investigated six times by the legislature during his tenure as state engineer, "usually prompted by some member whose pocketbook was being tampered with," Gilchrist explained, and each time it would emerge even more respected than when they went in, always able to refute all charges with meticulous records, proofs of thrift, and transparent dealings. When the Fergusons finally resorted to trying to push through three bills that would reorganize three different agencies—the Board of Control; the Fish, Game, and Oyster Department; and of course, the Highway Commission—in an attempt to give the governor a great deal more appointive power for rewarding her cronies, the efforts floundered and failed.

Another relic of Miriam's gubernatorial regime was the business-as-usual assumption on the part of certain contractors who had previously used the Ferguson crony and bribery system to obtain the contracts they wanted. These people especially resented the continuation of the clean, upstanding commission. According to the grandson of Judge Ely, who remained commission chair throughout the first six months of Miriam's term, until June 1933, they could prove a dangerous crowd—sometimes going so far as to threaten the commissioners themselves with bodily harm when they refused bribes. Once again, a former Texas Ranger came to the rescue.

Texas Ranger Frank Hamer. (Courtesy of Texas Ranger Hall of Fame and Museum, Waco, Texas)

Frank Hamer served as Judge Ely's armed bodyguard, accompanying him on trips to sites and districts and stating the clear message that no traditional Ranger or Highway Patrol officer would uphold any corrupt Ferguson practices or permit violence to those fighting them.

Gibb Gilchrist's knack for making and keeping friends through the openness of his character also had a direct effect on the culture of the Highway Department. During these years, the Highway Department "family" was born. A cohesive attitude of pride between employees and departments began to develop—with perhaps the exception of the Maintenance Division's complaints that they often had to come in and fix the mistakes that the Engineering Division made—and the completion, in 1933, of the new eight-story art deco State Highway Department building across the street from the Capitol in Austin created a center, the beating heart of the operations, and helped foster a sense of sturdiness and collaboration that spread out from the headquarters' "rock core of integrity" to all the far-flung division offices across the state.

Multimodal, 1930s style. Railroad underpass in Montague County south of Bowie. (TxDOT Photo)

5 The Wolves Return

The Highway Department should not be a political issue. It can be prevented from becoming one by a frank declaration of every candidate for Governor that he will keep his hands off the organization.—Judge F. W. Fischer, *Dallas Morning News*, May 30, 1936

Judge Ely worked on me for twelve months until he sold me on the idea of preserving trees. I was like the bashful lover; when I did fall, I fell hard.—Gibb Gilchrist, on revising his opinion of roadside beautification and landscaping

The next few years saw phenomenal advances in the Texas Highway Department. To the relief of Good Roads advocates, the Ferguson duo did not seek reelection in 1934. With State Engineer Gibb Gilchrist at the helm alongside a strong, unified commission, the fight to defend the department seemingly over at last, and the Fergusons supposedly slinking into the sunset, the department's priorities took on fresh energy. The atmosphere of the new headquarters buzzed with plans for genuine, forward-thinking innovations (rather than phony get-rich schemes like the Invisible Track or double-treated surfaces), brand-new safety modifications, colossal bridges made with groundbreaking technologies, radical changes in philosophies of land use, and enough cash to carry out the work. The hovering threat of federal fund loss had apparently ended. President Franklin Roosevelt's New Deal jobs programs bulked up the flow of money into the state. Thirty-six-year-old James Allred, a passionate Roosevelt supporter who replaced Miriam Ferguson in the Governor's Mansion in January 1935 (and the second-youngest governor ever to serve Texas), seemed to nurse no greedy aspirations toward the Highway Department or its resources; his main platform was to endorse and promote social programs. Even during pinched times, in such a lively and disciplined climate, new concepts flourished.

"The triple underpass will be one of the most imposing sights in Dallas. It will be located at the front door of Dallas since it will be the entrance for Highway No. 1 into the city—the most heavily traveled highway in Texas" (Dallas Morning News, *August 20, 1935). This structural triangular arrangement is a civic accomplishment of engineering genius. The arched gateway, constructed of concrete with distinctive art deco–style detailing, opened to wide acclaim and created a commanding entrance to the city from the west and an impressive exit from the east. During the fervor of the 1936 Texas Centennial,* Morning News *publisher G. B. Dealey officially dedicated the structure by riding in the first car that passed through it. The underpass gained more notoriety due to its proximity to President John Kennedy's assassination in 1963. (TxDOT Photo)*

Despite the shadow of the Great Depression that still darkened the country, Texas was on its way to becoming a leader in transportation ingenuity. Good leadership on the Highway Commission helped this trend along.

Greenery, Beauty, and Safety

Judge Ely had first approached Gilchrist in 1929, asking him to consider preserving the trees that grew next to the roadsides during construction. This was not a plan Gilchrist accepted readily; the European practice of tree planting and nurturing along verges seemed a petty concern compared to the more demanding needs in building the roads themselves. But Judge Ely gradually wore him down—to the dismay of many contractors and laborers, who considered this newfound reverence for trees a serious workday nuisance. And soon, the changing structures of the roadways themselves altered Gilchrist's views. Improvements in alignment, curvature, and grade had already been evolving; location now played an even bigger role in road designs, making it possible to keep high fills and deep cuts to a minimum. This in turn kept the roads more closely platted to the natural surface of the ground. The low grade line permitted a crown wide enough to both serve traffic and leave room for further widening at a future date. From these new designs evolved the modern "section"—a term defining a length of highway. As Gilchrist later wrote, "There was little need for such development prior to that time because with narrow rights-of-way and badly eroded deep ditches, there was no place for anything but the strictest utility. With modern rights-of-way now acquired, with a minimum width of 100 feet, usually 120 feet or wider, there is an opportunity to preserve the natural vegetation on the roadsides."

This policy made Texas one of the chief pioneers of the nation in wayside improvement and conservation. Gilchrist began by conscripting members of the Highway Motor Patrol to scatter wildflower seeds and cordon off existing trees during highway maintenance and construction. Women's clubs across the state had already been hard at work crusading for prettier roads and even planting trees, flowers, and shrubs in some places. The aesthetic boon to drivers everywhere was obvious, a true boost for tourism and recreational use. The advantages to safety and erosion purposes were less obvious but far more crucial. In 1931, the department established the Bureau of Roadside Development to better organize this work. Only two other states—Michigan in 1928, followed by Oregon in 1929—had already founded comparable statewide landscape programs. Michigan was also home to the nation's first roadside park.

Gilchrist grew increasingly enthusiastic, consulting experts to study and draw up lists of which plants and trees would thrive best in the many diverse Texas regions (such as native flora), which ones would require too

Cool and refreshing, this pool offered travelers a restful respite on their journey through Woodville. (TxDOT Photo)

much care and husbandry (ornamental exotics), and which ones looked most natural to their terrain. Eventually he reformulated every aspect of highway care to include respect for the roads' immediate surroundings, as exemplified by the directive he issued to the Maintenance Division in 1934: "Promiscuous mowing of the right-of-way should be delayed until flower season is over." He also later proudly wrote of other improvements: "The heavy earth cuts which formerly made unsightly scars are now reshaped into smooth form and covered with a vegetative mat for utility and beauty. Millions of square yards of Bermuda grass sod, vines, and other ground covers have been planted, and by continuing this work we will have all the ugly scars along the roads covered with a beautiful green mat." These aids to erosion control profoundly reduced the workload of the Maintenance Division. The previous scalping and sculpting of the landscape now gave way to a more upholstered effect—far less alienating and more welcoming to drivers.

Highway Department employees also recognized the value of roadside improvement and hospitality, as demonstrated by an enterprising maintenance worker named R. E. Wingate. Wingate lived in Woodville in East Texas. Inspired by his own vision, he undertook to co-opt 3 acres of Texas Highway

Department land in nearby Newton County, close to the Texas-Louisiana border. There he built a park with two bathhouses and a swimming hole for the general public's use—the first of its kind anywhere in the state. Two thousand people came to the opening of this free "roadside park" in July 1930. Later, Wingate created another two parks: one at Jaspar in 1932 and one at Woodville in 1934.

He also tried to persuade the Highway Department to buy and preserve 65 acres of East Texas virgin longleaf pine forest. Sadly, he failed at this endeavor. Longleaf pine trees, which take 100 to 150 years to grow to full size and may live to the age of 500 years, are now almost extinct, thanks to the clear-cutting, burning, and lumber deforestation practices of the last two centuries. If Gibb Gilchrist or the Highway Commission had shared Wingate's foresight and agreed to his proposal, the Highway Department's contribution to conservation would have been unique.

As far as Gilchrist was concerned, the official record did not acknowledge Wingate's swimming hole and bathhouses as the first roadside park in Texas. That honor went to another contender in Fayette County, when a Highway Department section supervisor, William Pape Sr., built a set of picnic tables, benches, and hearths in a stand of nine beautiful live oaks located alongside Robinson's Creek, on a 1.3-acre tract beside State Highway 71 between Smithville and La Grange. The land had been donated by the Czech residents of the small town of Hostyn, solely to benefit the public. Although the creek bed was often dry and no swimming hole offered a chance to bathe, this inviting setting encouraged weary travelers to stop, relax, refresh themselves, have a cookout, or enjoy a sandwich—a novel opportunity on a long rural stretch. Thus was born a tradition that has inflected travel in Texas to the present day.

In 1935, Lyndon Johnson was placed in charge of the Texas branch of the National Youth Administration (NYA), a federal program created to provide jobs and vocational training to disadvantaged young people. His first thought was to work with the Texas Highway Department, which he admired for its practicality, political autonomy, and good budgetary management, in directing teams of young men and boys to build roadside parks. What muscle Johnson had resented expending for the Highway Department as a fifteen-year-old, he now enthusiastically flexed again through the arms of his youthful proxies. Immediately the department assigned to them the plans for a park on Williamson Creek, just south of Austin on the San Antonio Road. "Lyndon's boys built a very nice park," Gibb Gilchrist wrote later, "and as far as I can recall, that was the only time I saw Lyndon Johnson in connection with this work. . . . Recently I found that the road [location] had been changed and the roadside park abandoned. Just as well, I guess, so that it would not have to be made into a shrine." According to other authorities, however, the NYA under Lyndon Johnson's management proceeded to build

*Old Highway 1 southwest of
Greenville. (TxDOT Photo)*

*Split rough-hewn logs make
a beautiful picnic table near
Simms in Bowie County.
(TxDOT Photo)*

Evolution of the Roadside
Park/Picnic Area

Roadside park construction, introduced into Texas in the early 1930s, soon became so popular that by the mid-1930s more than five hundred new rest stops had opened across the state, and by the late 1940s there were nine hundred. With no air conditioning and long drives without interstate amenities, travelers found the parks and rest areas a comfortable to place to stop, eat, and relax before continuing their travels.

State Engineer Gibb Gilchrist made a policy of encouraging regional architecture in these havens, overlooks, and picnic areas. The structures were designed to reflect and blend in with the local scenery, such as pine-bowered terraces in the East Texas woods, with *faux bois* tables and benches of concrete cast to resemble the gnarled branches and trunks of the surrounding trees, and closer to the Mexican border, high-peaked palapas with roofs of thatch that sheltered travelers from the hot sun in arid South Texas. Many of the early parks and rest areas were constructed by the National Youth Administration or through the Works Progress Administration.

Maintenance on the roadside park on US 67 south of Glen Rose. (TxDOT Photo)

Family picnic in a roadside park. (TxDOT Photo)

A shady arbor houses a cistern providing fresh water for travelers. (TxDOT Photo)

Oil derrick-styled rest area near Tyler, 1967. (Jack Lewis/TxDOT)

El Capitan in the Guadalupe Mountains provides a rest, a great view, and a photo op. (TxDOT Photo)

Dogwoods and pine trees provide shade and beauty on US 69 between Rusk and Alto. (TxDOT Photo)

Mid-1950s stop on Route 66 near Alanreed. (TxDOT Photo)

The tepee roadside park is a landmark on the River Road on the Rio Grande. (Jack Lewis/ TxDOT)

This is a tradition that has continued, though today's modern rest areas area a far cry from those of Depression-era Texas. Today. TxDOT has spent almost $200 million in the past decade to create a new generation of rest areas. Safety is the most important consideration, and amenities such as air-conditioned rest rooms, twenty-four-hour operation, and separate car and truck parking areas are regular features to entice modern travelers to take a break while crossing Texas. Visitors also learn about the geography, attractions, and people of the area.

This modern-day rest area locate between Alpine and Marfa provides restroom amenities and a chance to glimpse the famous and mysterious Marfa Lights. (J. Griffis Smith/TxDOT)

This retro art deco design is emblematic of the new generation of rest stops. Full amenities, interesting local information, and roadside beauty provide a safe haven for motorists to take a break and enjoy local color. This safe rest area is located on the old Route 66 near Donley in the Panhandle. (J. Griffis Smith/TxDOT)

Safety rest areas are built to encourage travelers to stop and refresh during long driving trips. Local and regional information and architecture add character to the highway and inform visitors about local history. The rest area near Pyote, on I-20, is located near the site of a World War II–era bomber base, the Pyote Air Force Station, and contains displays and information about the base's history. (TxDOT Photo)

This information board is found in the rest area near Hillsboro, just up the road from Abbott, hometown of Willie Nelson. (TxDOT Photo)

Shotgun

Willie

With all the outlawry in Hill County history, it is not surprising that Willie Nelson's music would be identified as Outlaw Country. Together with such musicians as Waylon Jennings, Kris Kristofferson and Jerry Jeff Walker, Willie rebelled against the constraints and predictability of the Nashville sound and brought an outlaw spirit to country music. The rebellion was as much in their lifestyle as in their music. They were sometimes outrageous, often a few sheets to the wind, and always absolutely original. And at the center of it all was the Red Headed Stranger, Willie Hugh Nelson. Willie was born in the nearby Hill County town of Abbott. His mother left town shortly after he was born, his father a bit later and so he was raised by his grandparents. When he was five his grandfather gave him a guitar and taught him a few chords, at age seven he wrote his first song, at age ten he joined his first band. Soon he was one of the most recognized artists in the world of entertainment. His voice is like no other. Although he has perfect pitch, the sound is as if he were singing from inside a pickle barrel, his songs are raw, melodic, poetic and surprisingly wise. Perhaps it was his Hill County farm background that caused Willie, together with Neil Young and John Mellencamp, to organize Farm Aid Concerts. For more than two decades, funds raised have been devoted to helping farm families remain on their land. Willie has said: "The fight to save the family farm isn't just about farmers. It's about making sure that there is a safe and healthy food supply for us all. It's about jobs from Main Street to Wall Street. It's about a better America."

Willie Nelson and son Micah

Young men working for the
National Youth Administration
(NYA) were put to work on Texas
highways, such as this back-
sloping project on Highway 5
near Gainesville. The NYA was
a federal New Deal program to
provide jobs for young people
ages sixteen to twenty-five. In
Texas, the NYA administrator
was former congressional aide
Lyndon B. Johnson. (TxDOT
Photo)

a total of 135 parks by June 6, 1936, providing jobs for thirty-six hundred young people. The largest single act of environmental innovation, however, took place in 1933 when Gibb Gilchrist hired Jac Gubbels to change the look and function of Texas highways, a decision that soon prompted change in highway design all across the United States.

Jac Gubbels had many diverse projects and interests, among them a passion for building "more attractive, safer and convenient highways for less money by taking advantage of natural forces and native materials." Gilchrist had hired Gubbels straight from the City of Austin, where he had been designing parks and boulevards. Although Gilchrist, in his newfound environmental fervor, conceived his brilliant idea of bringing Gubbels into the department in 1933, a 1934 mandate to spend federal funds on roadside landscape improvements paved the way for even more aggressive action.

A finished highway should, Gubbels felt, be "in harmony with the sur-rounding landscape" and have no monotonous straight sections that could lull the driver into an unwary trance before then confronting him with sud-den "angular, stiff . . . sharp lines and corners." As well as creating a "mental hazard" for the driver, these designs flaws were unsightly, "a separate bleed-ing scar" gashing the bare countryside, fields, and pastures. Rather than reengineer existing roads, however, he proposed to widen them and then reshape their boundaries with the illusion of variety. Tall trees, low shrubs, an ever-shifting profile would hide the flatness and tedium of surrounding scenery. Disguising deep ditches and cuts, bridges and culverts, ugly bor-row pits, and other distractions with layered plantings would guide drivers' eyes through visual cues and keep them alert to their task, stimulating what Gubbels called "road focus." Gubbels composed a list of plants according to scale, explaining their use on highway landscape features: "The psycho-logical effect of planting on the driver is dependent directly on the height of

Jac Gubbels

Jac Gubbels's birth in Groningen, the Netherlands, in 1897 marked the beginnings of a lifetime's adventures. Before reaching the age of twenty, he had studied landscape architecture in Germany and then, during World War I, traveled to Sumatra, where at age nineteen he worked as a plantation locater. After the war he returned to Europe but was unable to resume his studies due to lack of money, so in 1922 he set sail for the United States. He first took a job with a New Jersey landscape firm, then went on to help build an alpine botanical garden in Michigan. A planning firm in Denver hired him next, but in 1927 he left Colorado, moved to Houston, and opened his own landscape design office. His first Texas jobs both had a historical purpose: using old military plans, he restored the San Jacinto Battlefield to its 1830s appearance, and he redesigned the grounds and gardens of Sam Houston's house in Huntsville. When the City of Austin hired him to create parks and boulevards, Gubbels advised the city to buy as much open green space as possible. He then set in motion and guided the purchase of land along Shoal Creek, Plum Creek, and in Zilker Park, all of which are in use today in Austin's greenbelt system.

Jac Gubbels. (TxDOT Photo)

vegetation. We need dwarf shrubs at wing walls, behind road signs, on esplanades and circles, and as a foreground for taller growths. We need medium shrubs at bridge heads and on the slopes of low fills. We need tall shrubs and small trees on the slopes of deep fills to level them up. Taller trees form the background of the roadside, and serve a useful purpose in showing the alignment of a curving road and in warning the driver of intersections."

A dangerous traffic circle in Waco, developed in the early 1930s, caused more accidents and deaths than almost any other travel route in the state. It lay at the center of a confluence of five traffic arteries, all curving around it as they radiated outward. Gubbels described its perils in his 1938 book, *American Highways and Roadsides*: "The highways approaching this circle from three directions came over an open, level country where the excellent roadbed and wide-straight right-of-way encouraged the most speedy driving, say from fifty to seventy miles an hour. The circle was constructed with the usual curb, and the ground surface, slightly raised, was sodded and

"You follow Highway 58, going north-east out of the city, and it is a good highway and new. Or was new that day we went up it. You look up the highway and it is straight for miles, coming at you, with the black line down the center coming at and at you, black and slick and tarry-shining against the white of the slab, and the heat dazzles up from the white slab so that only the black line is clear, coming at you with the whine of the tires, and if you don't quit staring at that line and don't take a few deep breaths and slap yourself hard on the back of the neck you'll hypnotize yourself and you'll come to just at the moment when the right front wheel hooks over into the black dirt shoulder off the slab, and you'll try to jerk her back on but you can't because the slab is high like a curb, and maybe you'll try to reach to turn off the ignition just as she starts the dive. But you won't make it, of course."

—From the opening paragraph of *All the King's Men* by Robert Penn Warren

planted with flowers and low shrubs. Soon after the work was completed, wrecks became a common occurrence. Drivers would come in at full speed, crash the curb, and land their cars, sometimes upside down, in the flowers and shrubbery." Gubbels's solution to this menace was to have the circle filled in to form a 6-foot-high mound with sod on the sides and shrubbery and flowers on the top. "The motorist, coming in over a level highways, sees the solid mountain rising directly in his path and is warned to slow down," he wrote. The number of accidents dropped to almost zero afterward, and no lives were lost for the next few years.

Gubbels realized how changing the landscape's contours with lines of windbreak trees, dense tree canopies, carefully placed rock formations, shrubbery, flower plantings, and even vines "unconsciously [controlled] the turn of the wheel and the foot-pressure on the accelerator." In a 1940 article, Gubbels described a test ride with a friend along a stretch of "modern scientific highway," during which he observed the psychological effects of trees and other natural features on the driver: "The speedometer was climbing again, this time to 65, as we descended a long slope and sped toward one of those hilltops over which the road seems to disappear completely . . . where death stalks the careless. There were no warning signs at the side of this road. But there was another warning in the trees arching over the roadway ahead, seeming to crowd in from both sides, though not really doing so. They made the road look narrow, made the driver feel that he was going to go through a tight place. I saw my companion's eyes come to rest on that narrow opening, and they never left it. The speedometer began to drop . . . down to 60, 55, 50, 45! By that time we had risen above the crest of the hill and could see the road. The bottle-neck was behind us. The speedometer began to climb. . . . 'It works!' I shouted."

Long before Ladybird Johnson began her campaign to beautify America, she received tutelage by driving through the results of Gubbels's skillful re-shaping of Texas roadways. Safety landscaping for him, as for her, included both beautification and conservation. By preserving existing trees, shrubs, and wildflowers inside the state right-of-way, the Texas highway system benefited through the cheapest improvements possible. Bermuda grass planted along the drainage ditches and shoulders prevented erosion. Forming road-side lakes by setting a culvert on the slope of a stream instead of at the bottom of the main channel, filling the old channel with an embankment, and leaving the water outlet high up on the side so that the embankment acted as a dam—these measures meant that semiarid regions could now furnish water to roadside tree and shrub plantings. Drought-resistant vegetation, cacti, cane, tamarisk, willow: such natives visually helped to break up the bleak monotony while holding the soil together.

Gubbels became nationally recognized as the foremost exemplar in highway safety and beautification. By combining the two concerns in his problem-solving strategies, he helped transform the look and efficiency of highways across the country. Nowhere was this more clearly demonstrated than in Texas. By 1940, the Texas Highway Department had planted 9,600 miles of highway, controlled erosion on 13,995 miles, installed 15,260 miles with "good or moderate cross-section," and acquired 119 miles of additional right-of-way for tree preservation. Gilchrist completely embraced Gubbels's design principles and later announced that he and the highway commissioners considered their new landscape policy "one of the principal achievements of the Department" during the Depression years. There were to be several other significant achievements before the decade ended.

Early center-striping operation. (TxDOT Photo)

Law and Order, Contractors, Bridges—Centennial!

The Highway Motor Patrol, formed in 1929 to oversee inspection and weigh stations, had by the early 1930s evolved into a police force who patrolled the highways enforcing all aspects of the law—from speeding to smuggling, kidnapping, and apprehending armed robbers. Because they rode on motorcycles, usually in pairs, troopers often drove criminals to jail in the lawbreakers' own cars, then returned later to retrieve the motorcycle left on the side of the road. When the Texas Department of Public Safety was established in 1935, the Highway Motor Patrol was transferred into that department and renamed the Texas Highway Patrol, one of the three arms of that agency. A number of the Texas Rangers who had either quit or been fired by Miriam Ferguson during her dispute with them joined the Highway Patrol. This list included Frank Hamer, who tracked Bonnie Parker and Clyde Barrow under its auspices for 102 days before fatally ambushing them. Another prominent Ranger, the legendary Lonewolf Gonzaullas, became superintendent of the Texas Department of Public Safety's new Bureau of Intelligence, turning it into one of the best crime laboratories in the United States.

A formation of the motorcycle patrol unit with the Texas Highway Patrol in front of the State Capitol. (Courtesy of Texas Ranger Hall of Fame and Museum, Waco, Texas)

Texas Ranger Lonewolf Gonzaullas coined the phrase "One Riot, One Ranger." (Courtesy of Texas Ranger Hall of Fame and Museum, Waco, Texas)

During this period of the Depression, the Highway Department's alliance with contractors vastly improved from the ragged, difficult relationship that had prevailed throughout the Fergusons' regime. Bidding was now handled fairly and scrupulously. Great collaborations developed between the contractors and the supervising district engineers, and the results still grace Texas counties today. Throughout the decade, Texas received nearly $1.5 billion in work relief money, 40 percent of which was spent on road projects. From 1935 through 1939, this amounted to about $100,000 a day, equal to roughly $1.714 million in today's currency. Working with the department was a profitable enterprise, and contractors' companies bloomed all over the state—often growing into family-owned operations passed down from one generation to another. To keep these small companies healthy and handy for quick service when it was needed, Gibb Gilchrist and, later (and even more vigorously), Dewitt Greer, developed and encouraged the "small package" system: projects were divided into sections, with a different local contractor bidding on the desired section, often one that fell in the contractor's own hometown or county district. In this way, the local economies were given a boost, and the natural materials and resources of the area—gravel pits, quarries, concrete plants, equipment, and so on—were then available for future jobs, when more construction, road widening, and various kinds of repair grew necessary.

In 1918, when the new Texas Highway Department hired a University of Texas classmate of then–state engineer George Duren as the first official state bridge engineer, George Grover Wickline was employed by the City of Dallas. Before that he had worked as the McClennan County engineer, after holding a procession of other railroad, highway, and bridge jobs that permitted him to gain a wide experience base. Wickline managed the Bridge Section from 1918 to 1928 (at which time the section became a full-fledged division), designing and standardizing the solid bridges that the department installed over the upcoming decades. He then headed the Bridge Division continuously for the next twenty-five years, except during his three-year leave of absence, from 1935 to 1938, when he oversaw the construction of the Port Arthur–Orange (or Rainbow) Bridge across the Neches River. In September 1935, the Highway Department and Jefferson County together agreed to select George Wickline as the engineer in charge of what was to become the tallest bridge in the state (rather than highest—a fine distinction of deck height versus top of the span).

Prior to the construction of the
Rainbow Bridge across the
Neches River, a ferry carried

George Wickline (left) at the
Rainbow Bridge construction
site with an unknown coworker.

The south approach
to the Rainbow Bridge
under construction.
(TxDOT Photo)

B. P. "Buck" Greenward
and Robert Lytton, both
assistant engineers
when the Rainbow
Bridge was built in the
mid-1930s, pose high
above the Neches. Six
workers lost their lives
during construction of
the bridge. (TxDOT
Photo)

Riveting gang and inspector during construction working high above the Neches River. (TxDOT Photo)

Part of the crew responsible for painting the Rainbow Bridge, turned silver for their efforts. (TxDOT Photo)

The bridge's federal allotment had a deadline of December 15, 1935; this meant that all the plans and specifications needed to be drawn and a contract awarded by that date. The firm of Ash, Howard, Needles, and Tammen of Kansas City, one of the four largest in the nation, hastened through the design work, preparing the first portion of the plans by November 12, 1935, so that bids could advertised as quickly as possible. On December 2, three days before the deadline, the Union Bridge and Construction Co. of Kansas City, Missouri, was awarded the substructure contract. Union Bridge subcontracted with Austin Bridge Co. to do the approach work. A contract to build the steel superstructure was awarded to Taylor-Fichter Steel Corporation of New York on January 19, 1936.

It was Wickline's intention to hire only Texas engineers to prove that the Texas Highway Department could complete such a massive project. Since this was the first time a steel cantilevered bridge with caisson pier foundations had ever been built in Texas, these engineers would have had to acquire their experience elsewhere—out of state or even out of the country. P. V. Pennybacker acted as construction engineer and Wickline's top assistant. On September 4, 1936, work began on the span superstructure.

On December 31, 1937, the north and south spans were joined in midriver. The paving was completed on April 2, 1938, and the bridge opened on September 8, 1938. It was dedicated as the Port Arthur–Orange Bridge and replaced the Dryden Ferry as part of the "Hug the Coast Highway" on State Highway 87—the second-highest bridge over navigable water in the country, next only to the San Francisco Bay Bridge, and the highest in the South at the time. The reinforced concrete in its piers was one of the first uses of that material in this country. During the two and a half years of construction, six men lost their lives.

When completed, the Rainbow Bridge was the tallest bridge in the South over a navigable stream, 177 feet above the water. The vertical clearance was designed to allow passage of, what at the time was, the tallest ship in the US Navy, the dirigible tender USS Patoka. The ship never passed under the bridge. (Jack Lewis/TxDOT)

A Highway Department truck stands watch over severe flooding in Wharton in June 1935. (TxDOT Photo)

The department also in the mid-1930s instituted a Sanitary Patrol that would, through the upcoming decades, provide weeks of summer employment to many high school students as well as permanent staff in the removal and disposal of dead animals. During a four-month survey in the summer of 1934, the department determined that approximately twenty-one large domestic or farm animals were killed by motor vehicles per day across the state; the list included hundreds of horses, cows, mules, sheep, goats, and hogs—146 hogs in one county alone in that four-month period—but did not include the other nonlivestock carcasses, wildlife such as deer, raccoons, opossums, armadillos, skunks, coyotes, and javelinas, that also littered the shoulders. It became the Highway Department's task to either bury or burn these roadkills so that Texas highways would not turn into charnel lanes of rotting creatures.

In 1936, an extraordinary six-month exhibition opened on the State Fairgrounds in Dallas. It celebrated the one-hundredth anniversary of Texas' independence from Mexico and welcomed not just Texas, not just the United States, but the entire planet to join in: the Texas Centennial Exposition and World's Fair. Fifty buildings were constructed on the Fairgrounds to demonstrate facets of the two central themes, History and Progress. These included entertainment, the arts, cultural diversity, and natural studies,

J ournalist and author Joe Holley recounted his experiences in the Texas Highway Department Sanitary Patrol:

The Labor Day weekend got me to thinking about my first real paying job and about my work partner for that summer, Rufus Forrest Cochrum. We worked for the state of Texas, Rufus and I, for what was then called the Texas Highway Department. Our job was to pick up dead animals and trash along I-35 between Waco and West. We had no union, but we certainly labored. Rufus was 52 that long-ago summer, an old man in the eyes of a callow 17-year-old. Slender, with a protruding Adam's apple and a receding chin, he wore baggy jeans and a floppy-brimmed straw hat that resembled my Aunt Eunice's when she worked in her flower garden. Fair-skinned, he buttoned the sleeves of his blue work shirt to the wrist, while I, eager for an impress-the-girls tan, wore sleeveless T-shirts. (I'm paying for my foolishness these many years later.)

Rufus and I were a smoothly functioning team. We started our day early, Rufus behind the wheel of our yellow converted gravel truck, me riding shotgun with my foot propped on the dashboard as we crept along the shoulder. We had our eyes peeled for a tell-tale lump on the horizon that marked a dog's demise. We didn't specialize. We picked up everything but skunks. Once we found a sheep.

One morning we spotted a little brown puppy lying beside the road. I got out, deposited the little carcass in the bed of the truck and we went on our way. An hour or so later, I found the little guy wagging his tail and sniffing the piles of garbage all around him. Just stunned, I suppose. Maybe the smell had brought him back to life.

He rode up front with Rufus and me the rest of the day, maybe the happiest day of the summer. We couldn't stop talking about it. That evening the puppy went home with another garbage guy, a Texas Aggie working that summer on another truck.

To begin the trash portion of our day, we stopped in the shade of an overpass, hopped out of the truck and wielded a file to sharpen the nail on the business end of our broomsticks. Rufus would drive away, and I would loop around my neck a thin strand of rope tied to a tow sack. I would begin trudging toward the truck, which Rufus would have parked a couple of miles up the road. Meanwhile, he was walking away from the truck.

Spearing soiled napkins, crumpled newspapers and all the drivers' detritus that settled hour after hour, day after day, into the dusty, parched grass along the interstate—this was 20 years before "Don't Mess with Texas"—I'd get to the truck, empty my sack and drive toward Rufus. We'd repeat the process, leap-frogging beneath the merciless sun until we drove to the county garbage dump at 5 in the afternoon. I loved that stifling, smelly dump, my ordeal's-end Mecca.

A failed blackland cotton farmer, Rufus had been walking the interstate for nine years. During our half-hour lunchtime under a tree or during the mornings while we drove along together looking for dogs, he often told me in his high-pitched drawl about his retirement plans, 13 years away. I had trouble imagining 13 more days, let alone 13 years.

I may have grumbled at the tedium and at set-in-his-ways Rufus, but after a while I grew to respect the man. I hated it when the grader driver, the Cat operator, the asphalt patching crew—the guys with the manlier jobs—called me Little Rufus. Now, I'm embarrassed that I was embarrassed.

The truth is I got something out of that summer job besides the $1.05 an hour and the George Hamilton caramel hue.

For one thing, I learned never to look down on anybody who works hard, no matter the job.

I learned to appreciate competence and craft. "Soulcraft," the writer Matthew B. Crawford calls it in a book about the intrinsic satisfaction of manual work. Rufus had perfected his craft, however humble, and he took pride in doing it right. He made sure that the highway department and Texas taxpayers got their money's worth every working day of his life.

To raise awareness of the Texas Centennial and promote Texas, Governor Allred visited Hollywood. Here he dines with movie stars Ginger Rogers and John Boles. (Courtesy of Texas State Library and Archives Commission)

Travel information center, Farwell, on the
Texas–New Mexico border. (TxDOT Photo)

Wichita Falls. (TxDOT Photo)

from an aquarium to the Streets of Paris district to the renowned burlesque
fan dancer Sally Rand to the Hall of Negro Life—the first recognition of black
culture ever featured at a world's fair—to one of the most popular attractions:
the Cavalcade of Texas, a historical pageant covering four centuries of Lone
Star State history. Partly in preparation for the enormous numbers of tourists
who would presumably be flooding the state, partly in commemoration of the
Centennial itself, the Highway Department "conceived the idea of inaugurat-
ing a service that would aid traffic coming to Texas." And Texas had plenty of
traffic of its own. By 1936, the state held 1,525,579 registered vehicles.

A series of thirteen "stations" or travel information centers, architec-
turally designed to mimic their local vernaculars, were erected that year at
main travel entrances to the state by the Highway Department. Construction for all of them came to a total of $16,129.59 (as usual, Gibb Gilchrist's
meticulous accounts stipulated their cost down to the last penny). Some,
such as the ones in East and Northeast Texas, resembled in style traditional
Colonial houses in miniature or log cabins built by pioneers; some, such as
the ones in South Texas or farther west, were encased in stone or adobe—
"materials suitable to their geographical regions"—and of course, all proudly displayed Texas flags mounted on poles at their driveways' entrances.
Specially trained employees (Texas A&M military upperclassmen with high
scholastic standing) manned the "information houses," handing out free
maps to auto tourists. Their other duties mainly consisted of transmitting
accurate road condition information, supplying motorists with lists of
camps endorsed by the State Department of Health, and providing data on
the physical geography of the different state sections.

Travel information center, Orange. (TxDOT Photo)

Travel information center, Glenrio, along Route 66 in Deaf Smith County. (TxDOT Photo)

An early log cabin structure in Marshall (East Texas) housed the information center where department personnel provided travelers with maps and literature about Texas. (TxDOT Photo)

The first automobile maps had been published in 1935; now the department came out with the 1936 Centennial Map, a four-color production complete with a highly patriotic song written by Gibb Gilchrist, "Texas over All," printed in one corner. "From June 12th to December 1st, 1936, 29,600 cars representing every state in the Union and several foreign countries stopped to get free information at these information offices," Gilchrist later observed. In addition, the Texas Good Roads Association published the first issue of their new periodical *Texas Parade Magazine* in June, just in time for the Centennial Exposition opening, promoting Texas with information about facilities, highway development, Texas Highway Department staff news, letters from out-of-state travelers commending Texas roads and the brand-new information centers, tourism tips, and perky feature articles on sights to see along the way.

The Centennial Exposition attracted 6,353,827 visitors, including President Franklin D. Roosevelt, who attended on the sixth day after the official opening. It cost around $25 million to mount, generated ten thousand new jobs and an estimated $50 million to aid the economy, and was generally deemed a success, despite the fact that visitors' head counts failed to reach the projected tally. Jac Gubbels and his teams had also made landscape preparations for the Texas Centennial Exposition in Dallas by installing a huge number of attractive ornamental plantings along the roadsides leading into the Fairgrounds, almost all of which unfortunately were later uprooted and stolen by citizens.

Judge Roy Bean (on horse-back) dispensed West Texas justice from the Jersey Lily. The original building has been converted to the travel information center in Langtry, near Del Rio. (TxDOT Photo)

The old Jersey Lily now serves as the modern-day travel information center in Langtry. (J. Griffis Smith/ TxDOT)

The hallmark of the travel information centers has always been friendly, informed, and helpful travel counselors. Today, that tradition continues with counselors dispensing road condition information, travel maps and literature, route advice, and information about local attractions, shopping, and dining options. (TxDOT Photo)

The modern travel information center in Laredo shows how the facilities have expanded much like the state's safety rest areas. There are twelve centers located at major highway entrances to Texas and one at the State Capitol complex in Austin. (Michael Amador/TxDOT)

The striking architecture of the Amarillo travel information center on I-40 has won awards for its design and artwork, but the friendly and helpful service from the employees is the hallmark of the modern travel information centers. (Stan Williams/TxDOT)

But that same year, just one month before the celebrations kindled into the full firework display, controversy and disaster struck the department yet again. When the youthful James Allred was campaigning to succeed Miriam Ferguson in the governor's office, his reputation as a reformer and a Good Roads advocate seemed a guarantee of highway safety and ethical funding conduct, just as Dan Moody's had when he had followed her first term. It was a reasonable assumption to make. After all, Moody supported Allred, and it had been Moody who cleaned the Ferguson-soiled department stable, appointed good commissioners, and still remained a strong friend to the goals of the Highway Department. But appearances are often misleading. James Allred's real priorities soon grew clear. When Judge Ely reached the end of his six-year term on the Highway Commission in 1935, the statewide body of highway employees wished for him to be reappointed. He had served staunchly and faithfully to improve the integrity of the department as well as the roads, defending its thrift and good judgment while battling special interests, and although a second six-year appointment for an individual commissioner usually lay counter to the better wisdom of many, Judge Ely made the exception. He was, in Gibb Gilchrist's laconic words,

The department pulled out all the stops for the 1936 Centennial highway map, including not only the usual road and geographical information but also historical information, tourist stops, and even the words to the state song, "Texas Our Texas." (Courtesy of Texas State Library and Archives Commission)

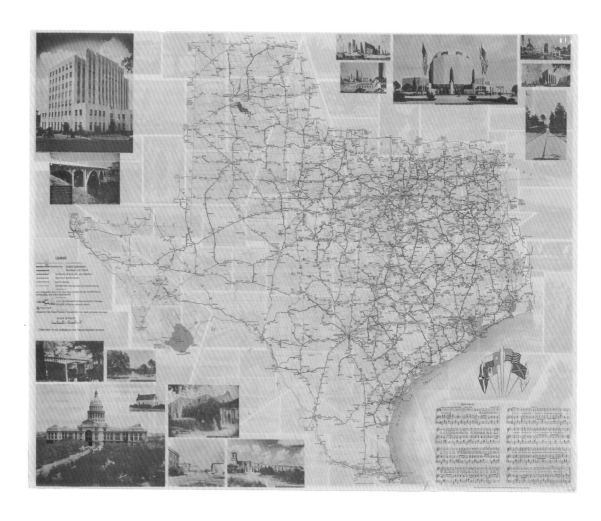

Division (district) engineers meeting in Tyler, 1936. Gibb Gilchrist is the tall gentleman in the center, and Dewitt C. Greer is second from the right. (TxDOT Photo)

The maintenance men of Walker County. (TxDOT Photo)

"popular and of such high character [that] we went along with that idea." Gilchrist secured a direct commitment from the two leaders of Allred's campaign (including Dan Moody) that if the entire department endorsed Allred and voted for him in the second primary run-off against his opponent, Judge Ely would be reappointed to the commission. Instead, soon after his inauguration, Governor Allred reneged on his pledge and appointed Harry Hines of Wichita Falls as Highway Commission chair.

Only a short time passed before Gilchrist and other department authorities came to regard this choice not only treacherous but calamitous. Allred's failure to keep his word seemed bad enough. But to place the officious Hines in a position of such power equaled a double betrayal. The political tentacles soon appeared. The stench of the old Ferguson crony era hung in the air when Hines called Gilchrist into his office one day and informed him, "From now on, I want every application for a job in the State Highway department anywhere in the state to be referred to me first, so that I may check it with the Governor's friends." Gilchrist was outraged. "I told Mr. Hines very bluntly that he was only one member of the State Highway Commission and that I knew the other two members would not subscribe to this to any degree and that I would do no such thing. He got red in the face, but that was the end of that." It may have been the end of that specific challenge. But far worse was coming. The ensuing mistrust was further aggravated by Allred's next raid on highway authority. As a determined apostle of President Roosevelt and the Recovery Act, Allred felt that social programs should take precedence over all other concerns as a cure for Depression woes and a protection for citizens. This philosophy included the powerful and popular movement, called the Townsend Plan, then captivating the whole country—the brainchild of a California real estate developer named Clements and his former employee, a sixty-six-year-old unemployed doctor and failed businessman named Francis Townsend. In 1935, President Roosevelt had embraced aspects of the Townsend Plan, revising its generous benefit scale downward to implement his own Old Age Assistance program for impoverished elderly people, paired with a national annuity that would later become known as Social Security. Governor James Allred apparently decided that now his turn had arrived to jump on the bandwagon and make a grand Texan gesture toward this movement. In May 1936, fifteen months after taking office and in the middle of making his latest campaign promises, he suddenly revealed his agenda: to extract $3 million from the designated highway funds to create old age pensions for Texans.

Then head of the National Youth Administration, Lyndon B. Johnson, visiting youth administration projects in 1936. Left to right: R. W. Jacobs, Lyndon B. Johnson, unidentified highway foreman. (LBJ Library)

To those in the Highway Department, it was if Allred had suddenly jerked a chicken out of a coin purse. The shock sent them reeling. Such a scheme would not only strip the department of carefully budgeted monies intended for ongoing jobs and projects, but it would also lead to a massive layoff: thousands of workers would lose their daily bread. Worst of all, the resultant shortfall would render Texas ineligible for government aid. Its matching federal funds would once again be cut off. And this time, no one could blame a Ferguson.

Fortunately, the other highway advocates in the state legislature stepped in and quickly rejected Allred's scheme. But his campaign platform attempt on the lifeblood of Texas highways had not merely disconcerted the department—it seemed a betrayal of all for which they had worked so hard. State Engineer Gibb Gilchrist especially was disgusted, and his disgust increased when Allred asked for his company one afternoon on an unexplained excursion to San Antonio. When they reached the scene of a large meeting, Gilchrist felt surprise.

There were some forty or fifty of Jimmy Allred's particular [influential] friends in San Antonio present. After our dinner, this meeting being in the early evening, they gathered around, and I soon learned just why Governor Allred had brought me to San Antonio. These men were seeking the designation of a state highway from San Antonio to Medina Lake [a beautiful recreational lake in the nearby Hill Country]. A good many speakers had something to say about it and, finally, the Governor took the floor and said something like this: "There is no need for any further discussion, because we have the man here who can accomplish this objective," and he presented me. I said . . . that I was not a member of the Highway Commission but only the State Highway Engineer, reminding them that a commissioner [Dock Martin] was in San Antonio but had probably not known of this meeting. [Martin had indeed not been invited or informed.] I told them that the usual procedure was that a delegation would come before the Highway Commission presenting claims for designations, and that the Commission would . . . advise them that they would have an investigation made of the claim [along with] a report. And that they usually instructed me to make the investigation and report. I told them that I wanted to be perfectly frank, though, and from what I knew and what I had heard, I saw no reason in the world why the State Highway Commission should designate a state highway from San Antonio to Medina Lake. That dropped the curtain on the whole thing, and Jimmy Allred didn't speak to me all the way back to Austin.

The situation could not continue. In the fall of 1936, James Allred was re-elected governor. By 1937, when he took the oath of office for the second time, the advances to the state transportation network looked quite impressive. The Texas Highway Department was now twenty years old. All twenty-six of the highway routes mapped out in its original plan two decades earlier were completed. Nearly 75 percent of these state highways were paved. The east-west Highway 1 from Texarkana on the northeastern state line, all the way to El Paso on the border—a monumental feat of con-

Early days at the Austin district headquarters at Camp Hubbard. (TxDOT Photo)

Maintenance warehouse in Jacksonville, 1937. (TxDOT Photo)

This style of bunkhouse was widely used by maintenance crews in the 1920s and 1930s. This one was still doing service as a warehouse in Sterling City in 1956. (TxDOT Photo)

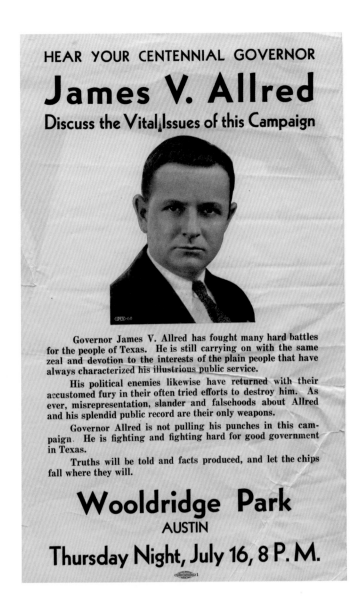

HEAR YOUR CENTENNIAL GOVERNOR

James V. Allred

Discuss the Vital Issues of this Campaign

Governor James V. Allred has fought many hard battles for the people of Texas. He is still carrying on with the same zeal and devotion to the interests of the plain people that have always characterized his illustrious public service.

His political enemies likewise have returned with their accustomed fury in their often tried efforts to destroy him. As ever, misrepresentation, slander and falsehoods about Allred and his splendid public record are their only weapons.

Governor Allred is not pulling his punches in this campaign. He is fighting and fighting hard for good government in Texas.

Truths will be told and facts produced, and let the chips fall where they will.

Wooldridge Park
AUSTIN
Thursday Night, July 16, 8 P. M.

(Courtesy of Texas State Library and Archives Commission.)

tinuous engineering covering 842 miles, and Texas' leg of the Bankhead Highway—now lay ready for use. So did its north-south counterpart, State Highway 2 (more or less the present route of I-35), which ran for about 520 miles from the Red River on the Oklahoma state line to Laredo, another city built at the boundary to Mexico.

The department continued to work on upgrades to improve traffic flow and many modern viaducts, overpasses, and underpasses. More than four thousand new bridges were under construction, and hundreds of old traditional bridges were modernized and reinforced. In addition, the legislature had established the state's first drivers' licensing—not only a fresh source of revenue but also a means of encouraging safe, responsible driving. These wonderful achievements had already transformed the lives, markets, economies, and industries in most of Texas. But approximately 14,000 miles of

The Bankhead Highway splits at this location: Highway 1 (now US 80) headed southwest via Ranger, Eastland, Baird, and Abilene; Highway 1A (US 80A) continued more westerly (what is now US Highway 181) via Breckenridge, Albany, and Abilene. (TxDOT Photo)

Highway 2 and US 81 in Williamson County, 1937. (TxDOT Photo)

state highway still remained to be paved—not even counting the many dirt county roadbeds—and some early-finished department-built roads already needed upgrades for heavier traffic and faster speed limits. The Materials and Tests Division that had been inaugurated by Gilchrist would soon become the model for the entire United States. The agency itself seemed to be in perpetual, progressive motion. And in the midst of all this activity, on September 1, 1937, Gibb Gilchrist resigned.

He had, as someone said, virtually created the department from scratch. He had built up its structure along well-engineered and well-organized lines. Under his guidance and policies, it had developed a healthier relationship with contractors across the state. He had defended it from looters, protected it against accusations posed by the unscrupulous, proven its excellent standards of work and product, its economy, its apolitical policy, its prudence. He had helped create a family, a system of collaboration between

By the end of the 1930s, car culture was in full swing in Fort Worth and other large Texas cities. (TxDOT Photo)

After a bridge washed out in 1935, the Highway Department operated this ferry across Lake Marble Falls in Burnet County for more than a year while the replacement was under construction. At first, passengers were expected to help hand-pull the ferry, but by the time this photo was taken, a modified 1929 Chevrolet was doing the hard work. (TxDOT Photo)

This "relief bridge" crossing the Neches between Woodville and Jasper in East Texas was built in 1939. (TxDOT Photo)

the private and public sectors, between employees communicating their unity across vast, open distances of Texas landscape. He had helped beautify roads and make them safer. Now he had reached a saturation point. The fight would always be ongoing; the revulsion toward political manipulation would only worsen or, at best, become a war of attrition; the unethical schemes of self-interested parties would, like the many-headed Hydra, spring up anew every year, for eternity. Such is human nature. "When I left Austin in 1937," Gilchrist said later, "I left governor's races over there."

By the time the Rainbow Bridge opened, Gibb Gilchrist had been named the new dean of engineering at Texas Agricultural and Mechanical College. Seven years later he would become president of A&M. In 1950 he conceived and oversaw the consolidation of Texas A&M's alliance with the Texas Highway Department, so specified in the original act, into a unique new institution, the Texas Transportation Institute (TTI), which developed into the first of two major scientific research and testing laboratories in Texas, unrivaled by any other like organization in the country. Meanwhile, his legacy at the Texas State Highway Department loomed so large that no immediate successor could have possibly met the raised level of employees' expectations, much less their hopes. The deep anger his departure stirred among the legions of people who worked there simmered on, reducing morale to its lowest ebb since the mid-1920s and resulting in hostility and belligerence toward his replacement.

Julian Montgomery, whom Gibb Gilchrist had known since Engineering School at the University of Texas and whom he considered a competent, reliable, "very able engineer," was no Gibb Gilchrist in the eyes of the department. He had first acted as State Engineer George Duren's assistant, back in 1917 during the department's founding. He had fought in World War I, returned to become district engineer for Corpus Christi, and then the Rockwall County engineer and city engineer for Wichita Falls. After that he opened his own private firm and was appointed regional supervisor for Texas and seven other states by the National Planning Board. Eventually, he was named director of the Public Works Agency for Texas in 1935 by Franklin Roosevelt—an appointment that probably appealed to James Allred's Roosevelt loyalties. In 1937 he resigned the directorship to accept the post of state engineer. But once reinstalled in the department, he was perceived as an outsider and, therefore, someone to be held in suspicion. He suffered the further disadvantage of being an appointee of Governor Allred's rather than a commission-selected and -hired state engineer.

Under Montgomery's aegis, the Texas Highway Department in 1938 broke with tradition and hired its first woman engineer, Leah Moncure, who the year before had graduated from the University of Texas to become the first licensed female engineer in the state. That same year, Montgomery put into operation a Workmen's Compensation Insurance Division and

During this 1938 inspection tour, a roadside park drew the interest of a herd of goats. (TxDOT Photo)

It took men, mules, and machines to get the job done near Tyler in 1938. (TxDOT Photo)

Laying down brick pavement on Highway 80 (today's Texas 180) between Weatherford and Mineral Wells. (TxDOT Photo)

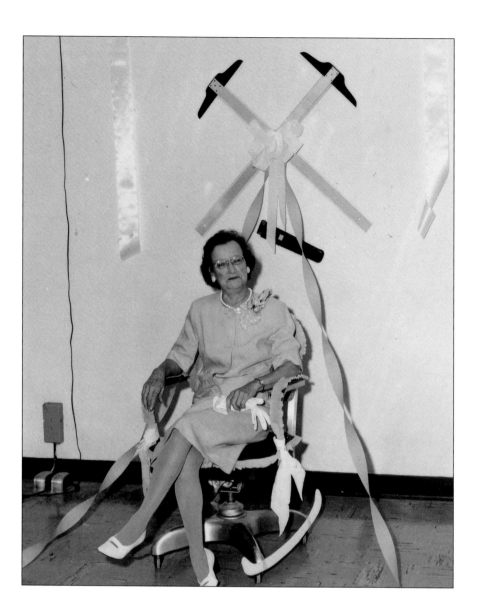

First woman engineer, Leah Moncure, at her retirement from the highway department in 1964. (TxDOT Photo)

established an employee safety program, with first-aid training and safety education. He also created the position of assistant state highway engineer in order to maintain closer contact between headquarters and the twenty-five far-flung districts.

That same November 1938, Miriam Ferguson entered one last race for the governor's seat, opposed chiefly by Texas Railroad Commissioner Renest O. Thompson and W. Lee "Pappy" O'Daniel—a folksy, colorful conservative Democrat. Pappy O'Daniel was a businessman who owned the Hillbilly Flour Company. A man of many interests, he also hosted a regular statewide noontime radio show, sang vocals with his band, the Light Crust Doughboys (later re-formed with different musicians as the Hillbilly Boys), launched the career of Western swing genius Bob Wills, composed songs to exalt his home state (including "Beautiful, Beautiful Texas"), claimed he ran his campaign on the platform of the Ten Commandments, and promised to

Despite being soaked by the rain, former flour salesman and music promoter W. Lee "Pappy" O'Daniel and his son and daughter put on their trademark exuberant campaigning for the folks. (Courtesy of Texas State Library and Archives Commission)

block a sales tax and raise pensions. The weary Ma Ferguson, who had once more against her will been persuaded by her ambitious husband to run, did not stand a chance.

Neither did State Engineer Julian Montgomery. He had lasted exactly twenty-one months as state engineer before retiring due to "health problems." Montgomery was finished once and for all with the arena of state-employed public service. But his successor, a short, slight, boyish young fellow from East Texas, would become a transportation giant.

Muddy roads were still abundant into the 1940s. (TxDOT Photo)

6 War and Peace

Roads are not like other defense weapons. They are not stored up in warehouses like guns and ammunition. They are not tied up at naval bases like fighting ships. They are not housed in hangars like airplanes. They are not held in waiting for the eventuality for which they were prepared.—Hal G. Sours, president of the American Road Builders Association, on passing of the Federal Road Act, September 12, 1940

I liked the whole idea of outdoor work, engineering and the status symbol that went with it: lace-up boots.—Dewitt Greer, on why he transferred from an electrical to a civil engineering major after a summer road construction job

My work is my hobby.—Dewitt Greer

When President Franklin D. Roosevelt signed the Hayden-Cartwright Federal Highway Act on September 12, 1940, he endorsed an older idea first explored in 1919 by the two US Army Transcontinental Transport Motor Convoys in 1919: American roads were crucial to national defense and needed to be well maintained and strengthened in readiness for bearing military personnel and equipment smoothly across the country in case of a possible invasion. No one in the military understood this better than General Dwight D. Eisenhower, the native Texan who, as a young lieutenant colonel, had himself helped man the first motor convoy across the country from Washington, D.C., to San Francisco. His precise and extensive log of the journey recorded the many vehicle rescues performed by Holt tractors, the wooden bridges splintered to bits by army machinery (eighty-eight), and the convoy's 231 "road incidents" (stops for adjustments, extrications, breakdowns, and accidents), as well as the nine destroyed vehicles and

The international bridge at Roma in Starr County, 1940. (TxDOT Photo)

Maintenance Foreman H. S. Byrd, 1940, recalled doing the dirty work: "I ruined a car or two and all the clothes I had in the world." (TxDOT Photo)

Underpass between Temple and Belton. (TxDOT Photo)

Highway 90 in San Antonio, intersection of Commerce Street and Zarzamora, circa 1940. (TxDOT Photo)

twenty-one men injured en route (out of 297) who were unable to complete the mission. The Washington–to–San Diego convoy traveling on the Bankhead Highway in 1920 fared little better, although it was smaller in size. Despite the modernizing of many highway systems throughout the states since those adventures more than twenty years before, they still fell far short of the condition necessary to support and expedite any military forces that would need to use them, should war ever be declared—a fact on which Eisenhower fixated his attention in years to come.

War was already razing parts of Europe. Throughout the late 1930s, the US Congress insisted on preserving American neutrality. Few people willingly conceded that US involvement on the European Continent might eventually become either recommended or necessary. But by the summer of 1940, after the defeat of France, the picture had changed. Great Britain now stood alone as the last democratic resistance to Germany, and the debate between intervention and nonintervention believers grew more heated. Isolationists still contended that the United States, half a globe away from the fighting, could avoid confrontation. Nor did many ordinary citizens imagine active warfare brewing on US horizons. President Roosevelt, however, was firmly convinced of war's inevitability, given the philosophy of dominance by force that the Axis powers espoused, coupled with their determination toward world control. To prepare for the potential threat, Congress and the president budgeted $163.5 million per year for a two-year plan to improve roads for military efficiency. Four days after signing the bill to effect this plan, President Roosevelt signed the Selective Training and Service Act, reinstating military conscription: the draft.

A total of 78,000 miles of highway were assessed to be of "prime military importance" by the authorities evaluating the national roads crossing the country. Of those, 6,375 miles lay in Texas. For reasons of direct territorial defense, the government plotted to renew the entire transportation network. And for those very reasons, the demographic map of Texas would change, too—irrevocably. Ironically, by the time the United States actually entered the war, so many funds had been diverted to national defense, weaponry,

Military maneuvers on a state highway east of Jasper. (TxDOT Photo)

equipment, and a well-supplied military that no money remained to supplement the road improvements previously deemed necessary. Neither were there enough laborers left in the country to carry out the work. Most able-bodied young men had either joined up or been drafted for military service—10 million inductees in all, out of 50 million men from ages eighteen to forty-five. This funds-and-labor dilemma did sow the seeds of highway planning for a future time when war would no longer be an immediate priority. But for now, what war was about to do to Texas, in terms of economics, industry, culture, population size, and shifting that population from a rural to an urban majority, would result in a more dramatic transformation than that occurring in almost any other state.

Exactly halfway through that same first year of the new decade, 1940, the Texas Highway Department was undergoing its own tidal shift. After resigning as state engineer (and apparently overcoming his health problems), Julian Montgomery moved into the private sector to head a group of architects and engineers. He made his own contributions to the war effort, some of which went far toward that dramatic statewide reshaping. Starting at the cusp of America's entry into World War II, he and his group of professionals spent the next several years designing and supervising the speedy building of many military bases, prison camps, and air force facilities, including Bergstrom Air Force Base in Austin. Meanwhile, the Highway Department needed a new boss, and rather go outside its ranks to search for one, the commission chose to promote from within.

At his birth, Dewitt Carlock Greer received his mother's maiden name tucked between his Christian and surnames. While he was growing up, his family and friends called him Dee. As a student in the Texas A&M Engineering Department, he was known as Cadet Greer and then Second Lieutenant Greer. But once he joined the Texas Highway Department in 1927, he became simply "The Kid."

This nickname, so resonant of a gunslinger's, was a nod to Greer's appearance: when he originally began his job as the city engineer for Athens, Texas, at the age of twenty-three, he looked so adolescent that he had been instructed to lose his Aggie belt buckle, grow a mustache, and start wearing a hat to counter his short, slender stature and boyish features—the effect of which, when as a district engineer he once bent to inspect a culvert-pouring project, prompted a contractor to say, "Move over, son, and let us get in here."

His childhood years took place in the tiny eastern town of Pittsburg, seat of the third-smallest county in Texas, located in the state's upper right shoulder. The middle son of a pharmaceutical salesman-turned-banker and a mother whose family had long been deeply embedded in East Texas, he began holding jobs early in his youth, daily living the values of his family by starting the morning with a paper route, sweeping out the bank after school, working part-time at the local drug store and the grocery store, at the box factory, in the peach-packing sheds, and the sweet potato sheds during harvest times, and elsewhere. The principles of frugality, self-reliance, and earning every penny while remaining debt-free that his parents instilled in him would stand him well for the rest of his life. At age sixteen he entered the Texas A&M Engineering School and graduated at twenty with a love for music and a strong-driving curiosity.

His postgraduation jobs went far toward informing his future in the Highway Department. Before the Athens position he held several others—the first for a contractor, filling out time sheets and payrolls, riding the rear end of an asphalt distributor while he controlled the burners and opened the valves, and operating a rock crusher. "I learned the hard way about building highways," he said of that employment, which stirred his doubt concerning the usefulness of his college degree. "I also learned how obnoxious a supervising engineer can be. This later proved to be of great value." From these manual tasks he moved on to briefly design mansion driveways and swimming pools for a landscaping firm in Dallas's wealthy Turtle Creek district. After that, as the landscape engineer of the new Texas State Parks Board, he spent a year camping out on the banks of the Guadalupe River near Boerne, supervising a team of twenty-six convicted murderers, rapists, thieves, and bootleggers while they prepared and developed the site of a state park. These felons had all been labeled trustworthy to work outside Huntsville State Penitentiary and were escorted to their new open-air

Some of the convicts to be supervised by Greer arrive at the Neff Camp. The Texas Ranger who delivered them is pictured second from the right. It is suspected that the woman in the front row is a nearby neighbor who often brought food to the camp. (D. C. Greer/TxDOT)

campsite tent homes in the Hill Country by a Texas Ranger who left them the next morning in the care of their unarmed twenty-two-year-old boss. After the Ranger departed, Greer told them, "You are honor convicts. If any of you want to leave, you can borrow my suitcase—but if they catch you, you'll be back behind the Walls."

Of that time, Greer later said, "It was the most interesting year of my life. I ate convict food, wore their clothes, slept in their tents . . . and never received a penny for it." The prison system supplied tents, equipment, and a convict cook who bought fresh fruit, eggs, chickens, and the occasional bottle of bootleg whiskey from neighboring farmers. They all worked six days a week, and on Sundays Greer permitted the inmates to swim, fish in the river, and "loaf around the camp." "The men liked the work," Greer said. "They were good workers. They enjoyed living in tents outdoors. Their ages ranged from the twenties to about forty-five." On Sunday afternoons, local families sometimes drove by to catch a glimpse of the criminals and show their children what such hardened cases looked like. "The men decided to put on a show for these people. They put the ugliest convict on a long chain, and tied him to a tree. When a car came by, he'd make a run toward it to the end of his chain. It would scare the hell out of the gawkers and tickle the convicts."

Occasionally the prisoners climbed into the camp truck so Greer could drive them to San Antonio, where they watched movies free of charge at certain city theaters. One night they refused to get back in the truck to return to life in the rough woods after the show. One spokesman stepped forward to explain: "Mr. Greer, we need to go to a whorehouse. We know some women at one here in San Antonio." Greer considered. Then he reached a conclusion that demonstrated his budding management skills. "I went with them. They didn't have any money, and they didn't cause any trouble. After a while, they all came back to the truck and we went back to the camp."

Governor Pat Neff had assigned Greer to this job before any appropriations were earmarked with which to pay his salary. By the time the year ended, Neff no longer occupied the governor's seat; Miriam Ferguson had supplanted him. In a move typical of her entire term, she methodically canceled the project: whether for financial compensation or through sheer liberality, she pardoned Greer's work crew, one man at a time. "I'd lend each one of them my suitcase as he went home," Greer said. "The last one kept the suitcase." Greer himself, left alone at last at the camp, gave a nearby rancher the camp's leftover foodstocks, shipped the equipment back to Huntsville, and borrowed $100 from a bank in Boerne with which to make his way back to Dallas and reapply for his former job with the landscaping firm. "Looking back, the idea [of prison laborers building parks] wasn't a bad idea," he reflected. "Afterwards, I corresponded with them for quite a few years. So far as I know, none of them ever went back to prison."

Greer was hired by the State Parks Board in 1924 to supervise convicts assigned to transform a rattle-snake-infested stretch of riverbank along the Guadalupe near Boerne into a state campground. The young A&M graduate, so youthful in appearance that he was mistaken for a student well into middle age, somehow executed the project with such tenacity, speed, and competence that all twenty-six of the convicts received pardons. (TxDOT Photo)

Greer's time with convicts from Huntsville was not the last time the Highway Department collaborated with the prison system. The License Plate Plant was constructed at the Huntsville Prison and opened in 1938. Prior to establishing this manufacturing system, the state bought license plates from contractors on the open market. In this photo, steel is being unloaded on the right, and prisoners are loading another truck with finished plates. (TxDOT Photo)

Brief History of License Plates

The first mechanism to fund state highways was vehicle registration. Before 1917, motorists paid registration to the county and were responsible for making their own license plates. Through the years, plate shapes, designs, and style have changed.

The earliest license plates (issued before the Highway Department was established) were homemade and came in a variety of shapes and styles. (Texas Department of Motor Vehicles)

First state-issued plates and registration medallion, 1917. (Texas Department of Motor Vehicles)

Vehicle registration medallion from 1918 (TxDOT Photo)

License plate from 1929. (TxDOT Photo)

Commercial plate from 1942. (TxDOT Photo)

Shirley Battey shows off the new 1955 license plates. (TxDOT Photo)

Robert Townsley (right) directed the department's Vehicle Titles and Registration Division from 1960 to 1986. Here he presents Governor John B. Connally with a set of personalized plates. (TxDOT Photo)

Two special-edition plates, the 1968 Hemisfair and 1936 Centennial license plates, are displayed by employee Julia Reig. (TxDOT Photo)

Robert W. Townsley gives the first special license plates to Mexican consul Rafael Linares in 1965. (TxDOT Photo)

Congressman Jim Wright and Commissioner Reagan Houston are putting special Texas license plates on a lunar rover in 1973. (TxDOT Photo)

One of the last embossed license plate designs displayed a number of Texas icons, including a cowboy on horseback and the Space Shuttle. (Texas Department of Motor Vehicles)

The current general-issue license plate design returned to basic black and white, preferred by law enforcement. The state offers a multitude of optional designs through the organizational plate program and a third-party vendor who sells premium designs. (Texas Department of Motor Vehicles)

The tastes for hard construction labor, negotiating, and getting along with contractors; the penalties of bad supervision; the pleasures of problems elegantly solved, of good design well applied; the management of "difficult" men: all these factors contributed toward Dewitt Greer's suitability for a meteoric promotion. "He was interested in people," said a colleague. "He got into all phases of operation for his own edification." Once he joined the department as an assistant resident engineer and "instrument man," laying out a highway system for Henderson County, State Engineer Gibb Gilchrist quickly recognized his talents, lack of egotism, and practicality, and two years later placed him as district engineer in Tyler. Then, over the next nine years, Greer absorbed Gilchrist's mentoring: "Gibb Gilchrist was probably the guiding light of my life. . . . [He] is the man who set the pattern of honesty, integrity, and hard work that molded the department. Gilchrist was tough, but he was the best man to handle the transition from corrupt politics. He made the way easy for the rest of us." By 1936, Gilchrist had promoted him to head the department's division of construction and design—the youngest person ever to receive that appointment. He and his wife and daughter moved to Austin. His younger brother Robert, now a civil engineer also, had followed in his brother's professional path and in 1935 had taken a job as the Texas Highway Department's junior resident highway engineer in Dallas. This fit with what would become a traditional pattern; through decades to come, the Texas Highway Department fostered a family atmosphere in more than mere camaraderie and collegial relationships. Entire generational tiers of relatives collaborated to develop roads and bridges; fathers, sons, and grandsons, as well as nephews, sisters, and cousins, sought to contribute public service to Texas through the department and helped weld it into the steel-solid structure it became.

The new state highway engineer, D. C. Greer, got his first look at all of his team leaders at the district engineers' meeting in November 1940. Greer is on the front row (far left) in front of the State Highway Building that would one day bear his name. (TxDOT Photo)

When, in 1937, battered and disgusted by his recent battle with Governor James Allred, Gilchrist left the department for College Station, his protégé Greer remained. Then, in 1939, a tragedy occurred in the Greer family that would forever change Dewitt Greer's views toward highway building—in particular, road safety and its applications: his mother was killed by a drunk driver in a car accident in West Texas. When he replaced Julian Montgomery on July 1, 1940, Greer's dedication to policies of safe driving and safe conditions grew paramount in all his future endeavors. Directing the department's engineers to "constantly strive to design accident-proof highways" and to "design for safer accidents," he continued the philosophy that Gibb Gilchrist had espoused by hiring Jac Gubbels to create safety landscaping and led the way toward higher safety consciousness as the department's first priority. At the time of his appointment, his name changed again—this time for good. Although still boyish and jaunty-looking, he was "The Kid" no longer: now and everlastingly, he would be known as "Mr. Greer."

Greer (third from left), Highway Commissioner Reuben Williams (on Greer's left), a county judge, and four county commissioners inspect a highway project in the summer of 1941. (TxDOT Photo)

No longer "The Kid," Greer and W. J. Van London, the engineer-manager of the Houston Urban Project, team up for an inspection in 1945. The project group lasted until 1984 and constructed Highway 610 (the Loop) and all its associated freeways. (TxDOT Photo)

War: Waiting, Woe, and Willpower

The growing trend of family service in the Texas Highway Department met with a challenge immediately after Greer took charge. His brother Robert, by then a five-year employee, symbolized the possibility of an illegal scandal in the form of nepotism. Attorney General Gerald Mann reviewed the case, rightly concluding that no favoritism or nepotism could have taken place, as Dewitt Greer had not held a position of hiring or firing anyone outside his own district when Robert joined the department. Not long after, the issue became moot; Robert left to pursue military service and was assigned the post of captain of engineers at Corregidor in the Philippines.

Many other department employees also began resigning their jobs to enlist in the military, even though the government had designated the agency as essential, and therefore its personnel was not subject to the draft. A total of about eighteen hundred men and women, or 30 percent of the department's workforce, wound up performing war service. In 1942 Greer himself wished to join them by entering the navy, but the US naval authorities refused him, regarding his current placement as vital to national defense. He made a promise: The department, he said, would tolerate no draft dodgers. But all those who resigned for patriotic action would have their

Even during war, the opening of a new bridge or highway would often be a time of celebration for communities. Austinites gather for the opening of the Lamar Street Bridge in 1942. (TxDOT Photo)

jobs waiting for them when they came home again. Already he had been cultivating grand ideas for excellent, low-cost highway expansion. The government's list of military roads had looked to be a golden opportunity. But when the Japanese attacked Pearl Harbor and Corregidor and the resources evaporated, Greer accepted that for the next few years, a quietus would fall on progress. Besides, another element had been added to his list of concerns: the Japanese had now captured Robert, holding him prisoner in a camp in the Philippines.

- -

Military presence dominated wartime Texas. The number of soldiers, sailors, marines, pilots, WAVES (Women Appointed for Voluntary Emergency Service), WAFs (Women in the Air Force), and WACs (Women's Army Corps) exceeded that of any other state, reaching more than 750,000. Before 1941, according to census figures, more people lived in New York City than in the entire state of Texas; its population accounted for only 5 percent of the nation. Yet 7 percent of the recruits that constituted the American armed forces were Texans, including such major leaders as General Dwight David Eisenhower, Admiral Chester William Nimitz, and Colonel Oveta Culp Hobby, the first director of the WAC as well as the wife of former governor William P. Hobby, who had replaced the impeached James Ferguson and then been elected in his own right. The local population numbers changed dramatically also; throughout the war, 1.5 million military troops were stationed in Texas for training. Many civilians moved here to work in war industries, supplying troops overseas with Texas oil and fuel, beef, medical supplies, weapons, and other kinds of equipment. This helped increase the net population by 33 percent, forever shifting the balance of residents, both old and new, from rural districts to cities. When food prices climbed, farmers planted crops to the absolute limit of their soil's capacity, thereby providing grain to the Allied countries. Manufacturing quadrupled. The steel mills in Houston and Daingerfield drew both out-of-state newcomers and former agricultural dwellers. Texas City became the home of the largest smelting plant in the world.

Garland, Fort Worth, and Grand Prairie mushroomed when gigantic aircraft factories covered previously empty fields and pastures. The vast shipyards filling the docks of Beaumont, Corpus Christi, Galveston, and Houston turned those cities from calm commercial ports to bustling military centers almost overnight. In East Texas, the paper and pulpwood industry revived in various places throughout the second-generation pine forests. Elsewhere in the state, clusters of factories produced synthetic rubber and munitions. Just as important to the nation and the war effort were the Inch Pipelines—the Big Inch and the Little Big Inch—the longest pipe-

D. C. Greer pitching horseshoes on a ranch in Kerr County in 1942 on "one of the few vacations I've ever taken because I travel so much anyway." (TxDOT Photo)

lines in the world at the time, conceived in 1940 by Secretary of the Interior Harold Ickes and built in 1942 and 1944 as emergency measures to flow oil overland from Texas to New Jersey. They became critical to the country's defense after German U-boats attacked transshipment oil tankers full of Texas oil along the Eastern Seaboard and Atlantic shipping routes. The fact that they are still in use today amply demonstrates both their efficiency and the roles that Texas products were playing during the war's duration.

More than sixty base and branch prisoner-of-war camps were established in small towns and in the countryside—more than in any other state—and the three internment camps, in Crystal City, Seagoville, and Kenedy, where both American and foreign individuals from Europe and elsewhere (mostly of Axis nationalities or origins: German, Italian, and Japanese) suspected of being security threats were detained by the Immigration Service. Altogether, the state welcomed 175 major military installations, plus numerous minor ones, including 65 army airfields, 35 army forts and camps, and 7 naval stations and bases.

But the roads and bridges remained subpar for the loads they had to sustain. A graphic example of this problem was an incident that took place in Somervell County in mid-January 1945, when an overloaded army truck tried to cross the Texas Highway 67 bridge over the Brazos River. The attempt resulted in the total destruction of the old 1908 structure. One soldier was killed, and two more were gravely injured. This was by no means an isolated occasion.

Controlling erosion with sandbags in the Tyler District, 1945. (TxDOT Photo)

A US Army truck known as a prime mover transports a "Long Tom" 240 mm artillery piece to Fort Hood in 1945. (TxDOT Photo)

A Douglas DC-3 is transported along Ledbetter Drive in Dallas. (TxDOT Photo)

Airplanes being transported in Kaufman County. (TxDOT Photo)

Maintenance of the existing roads was extremely limited due to gasoline rationing, shortages of asphalt and tar, and, of course, a scarcity of people to shovel and dispense them. New construction, along with federal highway matching funds, was out of the question. The only step Greer could take toward the future was to invest all state-generated revenues from registrations, drivers' licenses, and the gas tax in short-term government securities so they could be protected from other political attempts to nab their use and earn an increase during the shielding. (Fortunately, neither the colorful and boisterous Governor Pappy O'Daniel nor his highly conservative, laconic successor, Coke Stevenson, tried to suborn state funds for his own agenda; in fact, Stevenson was credited with the post-Depression economic recovery of wartime Texas during his long stint as governor, from 1941 to1947.) The remaining staff Greer instructed to start preparing detailed plans for conducting the enormous highway expansion, once peace was restored, that he had already envisioned. Some of these included a design for expanding Dallas's Central Boulevard (which is now the twice-revised Central Expressway), an outer beltway loop around Dallas (Texas State Highway Loop 12, since superseded by Interstate 635, and after that, the President George Bush Turnpike, otherwise known as State Highway 190), and a major Dallas street remapping.

When at last the conflict ended and the battle casualties were reckoned, the losses dampened the joy of many department employees' returns. Twenty-two of their workmates had met with death in action, including two missing and three taken prisoner. Among these, the heartbreaking fate of engineer Captain Robert Greer would haunt his brother for the rest of his life. Robert had been shipped from a prison camp in Mindanao in the Philippines to incarceration in Japan and for nearly three years had survived ordeals and privations. During that period, starting one year after his disappearance, his wife and his father received several postcards from Mindanao Prison Camps One and Two, each card carrying an almost identically worded message: "I am well and strong and everything is all right. Love to everyone." Finally, in February 1945, six months before Japan surrendered to the Allies, he died—it was suspected through starvation and mistreatment. He left a widow and a six-year-old son who had not seen his father since the age of two. Robert was thirty years old.

Peace

Now that the war was over, Dewitt Greer plunged all of Texas into full-steam-ahead mode for highway construction. Having saved up the resources toward implementing his visionary plans to help create what he called "a civilization geared to motor vehicles," he was already a stage beyond other states, both fiscally and practically. Thus, in 1947, less than one and a half

A "blow up" takes out Highway 71 north of El Campo in 1947. (TxDOT Photo)

A weakened bridge in Zavalla County. (TxDOT Photo)

years following the war's conclusion, Texas accounted for 25 percent of the entire nation's road construction. "The highway system of 1946 was a casualty of World War II. The roads in place were deemed expendable to the war effort," he said, as he set about rectifying that situation.

Unfortunately, the new construction still left some people disadvantaged. Despite tremendous progress, first under Gilchrist and now Greer, the pledge of the Good Roads Movement, to "get the farmer out of the mud," had not yet been fulfilled. Many rural dwellers remained as isolated, inaccessible, and marooned as they had been before the Texas Highway Department's birth, forced to rely on makeshift paths, weather-dependent roads, and rutted dirt tracks. Yet an innovation loomed on the horizon—a scheme that no other state had yet conceived, much less instituted.

Bullet-riddled sign marking the first farm-to-market road, FM 1. (James Cammack Photo)

Although, technically speaking, the first farm-to-market road had been completed in 1937 in Rusk County, solely in order to link two tiny communities 6 miles apart—Mount Enterprise and the now-extinct Shiloh—the designation was not yet official. (That original route is now part of Texas State Highway 315.) The first certified, paved, two-lane farm-to-market road, FM 1, was constructed in 1941, also in East Texas. It connected US 96 to a sawmill belonging to the Temple Lumber Company.

Throughout the war, no further progress was made. Then, in 1945, the Highway Commission authorized a three-year pilot program for building and paving a 7,205-mile, two-lane system of country roads, the ones netting half of the state east of what is now US 281, or in some places Interstate 35, to be labeled farm-to-market roads, the ones laced across the western half to be called ranch-to-market. (A few exceptions to this rule included Henderson County in East Texas, as well as Comanche, Erath, Reeves, Brown, and several other West Texas counties, where the rule would be reversed.) Immediately a battle broke out in the legislature. Lobbyists vying for road monies to go instead toward large arterial routes stymied the program, begrudging even the matching federal funds that would help pay for it, just as they had fought for urban-versus-country road dollars in the 1920s and 1930s.

Then, in 1949, the legislature took a fresh step toward the problem: it passed the Colson-Briscoe Act, sponsored by State Representative Dolph Briscoe and State Senator Neveille Colson, which established and guaranteed permanent funding to connect the vast, empty spaces in the Texas landscape and end the seclusion of rural peoples for good. This act appropriated funding for a very extensive infrastructure of secondary roads, hiking the gasoline tax by one cent per gallon of that portion dedicated to local highway construction, plus an annual dedicated reserve of $15 million (which in 1962 was increased to a yearly sum of $23 million with matching federal funds, along with an expansion of the system from 35,000 to 50,000 miles). Roads that had been previously maintained by counties would now join the other county roads that had become the Texas Highway Department's responsibility, thus alleviating the counties' burden. Farmers and ranchers could travel into town to sell their products and purchase goods without the laborious, time-consuming obstacles that had always prevented easy movement; their children could ride to school on buses without missing days and weeks due to stormy weather; rural mail delivery would not now encounter its previous impediments. At last the farmer's mud extraction dream had become a reality. The fact that one of the bill's two sponsors—Senator Colson—was a woman added even more to its significance: until this moment, few woman had had much opportunity to contribute to the transportations system in Texas.

Mules pulling a car out of the mud was still a common site on Texas rural roads at the end of World War II. This road in Leon County near Oakwood is now FM 542. (TxDOT Photo)

From Farm to Market

Farm-to-market and ranch-to-market roads are a uniquely Texan innovation. Although Missouri started a similar scheme in 1952 called supplemental routes, and Iowa has a farm road network (under county rather than state jurisdictions) that is eligible for state aid and receives occasional shots of federal fiscal aid, no other state has ever built such a state-sponsored and -maintained transregional system. Not even California has a farm-to-market road program, although it contributes half of the agricultural produce consumed in the United States; only now is it starting to study the conditions to develop one.

Currently the farm-to-market/ranch-to-market road system comprises over half of all the mileage under the TxDOT aegis, although a number of these roads trace through areas that have now evolved from pastoral, agrarian environments to heavily trafficked cities and suburban areas—such as Westheimer Road, a 19-mile six-lane street, stretching through strip malls, fast-food chains, gargantuan apartment complexes, and subdivisions, deep into the heart of Houston's most expensive retail district and wealthiest neighborhoods. In 1995, TxDOT made an attempt to publicly

The farm-to-market, ranch-to-market, and ranch road program connected mostly rural areas with access to markets and commerce. But the programs also created some of Texas' most beautiful drives. A good example is RR 337, which winds through the Texas Hill Country from Camp Wood to Medina. It was first designated in 1945 and built in segments over the next three decades. (Jack Lewis/ TxDOT)

Putting up the sign at the intersection of State Highway 73 and FM 563 in Anahuac. (TxDOT Photo)

RR 337 road sign. (Jack Lewis/TxDOT)

rename some of the city-engulfed roads with an alternative title: urban roads. Ironically, the people most opposed to this plan were local urban residents, who protested the new term as "un-Texan." The cost of changing the signage was also considered prohibitive, so TxDOT simply redesignated these particular roads as "urban" for its own classification purposes within the department. Nowadays, although the state continues to maintain and call them farm-to-market, it does not provide funding for their expansion—leaving that task and its costs to the cities.

FM 170, the River Road, stretches along the Rio Grande from Study Butte to Presidio. Though it provides a direct link for goods and services along the remote Texas-Mexico border, the road is also a haven for recreational travelers and offers unparalleled views of scenes of wild Texas. (J. Griffis Smith/TxDOT)

Meanwhile, the department began to thrive in another way: well-trained and well-qualified veterans were returning to civilian life, many of them looking for stable jobs in their fields. Included were engineers who had gained experience through rigorous circumstances, meeting difficult problems overseas. A number of these new employees would go on to spend their entire careers in the Texas Highway Department and make major contributions to the transportation system—employees such as Burton Clifton and Marquis G. Goode Jr., future executive director, who had been drafted by the army halfway through his Texas A&M engineering course to build bridges in Europe and Japan, come home to complete his degree, and then rejected several more lucrative job offers in the private sector to work for the state.

By the time the next decade was a couple of years old, more than 24,000 miles of country roads had been added to the state's system, improved with "hot topping" (asphalt mixed with gravel) to defy bad weather and heavily loaded vehicle wear. The plans Dewitt Greer had laid, the state highways for which he had assumed the burden of responsibility, had also now been methodically transformed into 22,000 miles of excellent pavement, unrivaled

The road system in Texas is constantly evolving; here, the old meets the new as an archaic wooden bridge is being replaced by new construction in a scene from 1949. (TxDOT Photo)

anywhere else in the United States. In 1948, the Houston Freeway (US 75) opened, running from Houston's northwest border through downtown to eventually connect with the Gulf Freeway, also a segment of US 75, which had started constructed in 1947 under the direction of Houston's urban supervising engineer J. C. Dingwall as Texas' first urban expressway. One of the features of the Gulf Freeway that made it unique among other similar US thoroughfares was the creation of continuous and contiguous frontage roads that offered ready access to the neighborhoods and businesses on either side, increasing the economic health and square-foot real estate value of these areas—a design factor that Greer included in his plans for the interstates built during the following decades. Other expressways on the Highway Department's drawing boards quickly followed. After the Houston-to-Galveston Gulf Freeway opened in 1951, an average of sixty-nine thousand vehicles per day traveled on it during its first year—a sum that amounted to forty thousand more than that route's 1948 level of use. Clearly, Greer's projects were getting implemented not a moment too soon.

But a sad human story lay behind some of these innovations. The new 300-foot rights-of-way with their two-way frontage roads on both sides of the Houston Expressway set an additional precedent for what would become a regular practice of planning and building cross-city routes: they cut through decrepit or dilapidated neighborhoods conveniently labeled "urban blight" by the authorities and therefore considered not only expendable but desirably so. This policy led to the destruction of homes and buildings (some of historic significance) and displacement of people who had often lived there for generations. The same thing occurred in Dallas when the first iteration of Central Expressway came under construction in the 1940s. The initial 2-mile length that sliced through Dallas prompted the flattening and erasure of several venerable African American neighborhood communities of nearly a century-long occupancy, rendering homeless more than fifteen hundred residents. One traditional institution to suffer was the Freedman's Cemetery. Unbeknown to the planners and engineers at the Texas Highway Department, an entire quarter of the 4-acre cemetery lay on the route's southern end, and the graves of over eleven hundred African American men, women, and children dating back to Emancipation were paved over with asphalt and concrete to smooth the way for midcentury urban traffic—a fact that remained undiscovered until renovations of the route took place decades later.

Sections of Central Expressway opened in 1950. By the end of 1952, the stretch from downtown to Mockingbird Lane was functional, and the rest of the route, all the way out to Campbell Road in Richardson, was completed and opened in August 1956—just in time for all the authorities involved to realize that North Dallas and Richardson had already expanded beyond what anyone had foreseen.

Evolution of North Central Expressway

Looking north at US 75, just south of Hall Street in Dallas, late 1940s. (TxDOT Photo)

ARCHITECTURAL STUDIES OF PROPOSED EXPRESSWAY CONSTRUCTION

Motorist's View of Overpass at Ross Avenue

Aerial View of Overpass at Mockingbird Lane

Aerial View of Underpass At Knox Street

Typical Perspective Section of Expressway at Grade

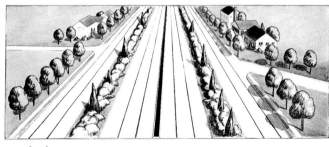

Page 5

An early postwar study commissioned by D. C. Greer about the development and design of Central Expressway, 1946. (TxDOT Photo)

An early view of US 75 and the Loop 12 overpass. (TxDOT Photo)

From its beginning, Central Expressway proved to be a crucial and much-used conduit through the growing metropolis of Dallas. By the late 1980s it had become apparent that an expansion was necessary to accommodate the increased usage. (TxDOT Photo)

The plan actualized in concrete and asphalt in 1962. The Dallas skyline was growing, and Central Expressway served greater traffic demands. (TxDOT Photo)

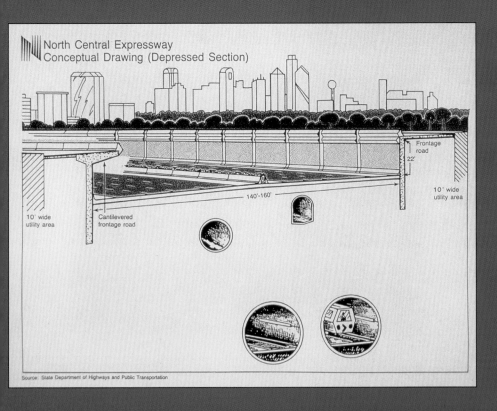

North Central Expressway
Conceptual Drawing (Depressed Section)

10' wide utility area

Cantilevered frontage road

140'-160'

Frontage road 22'

10' wide utility area

Source: State Department of Highways and Public Transportation

The solutions to widening an essential artery with such highly restricted right-of-way depended on two possibilities: either rising upward or going down. The federal authorities in charge of the funding needed to help implement the expansion required that the first option be selected. But the higher-end neighborhoods located along Central protested such an unsightly and intrusive solution, and the decision was made to dig and depress the roadway instead—a far costlier project that prompted the withdrawal of federal aid and relied solely on state monies. Within the right-of-way, two tunnels were required, one for the Dallas Area Rapid Transit route to downtown and one to carry water away from the depressed roadway segments. (TxDOT Photo)

With the reconstruction of North Central, massive underground tunnels and holding areas were required to remove rainwater from the depressed section of freeway. (TxDOT Photo)

The amount of infrastructure already
present and buried under the expressway
was enormous and necessitated removal
and replacement of power cables, communication lines, fiber-optic cables, and
train tracks as part of the deepening and
excavation process, as well as the digging
of vast tunnels to divert floodwaters from
the walled canyons of the new roadbed.
The waters were shunted through these
tunnels into storage caverns underneath
a city park that now memorializes the
Freedman's Cemetery, the oldest cemetery in Dallas, and one in which the graves
of more than two thousand freed African
American slaves and their descendants
were displaced when the first iteration of
Central Expressway was carved out of
the old Freedman's Town neighborhood.
(TxDOT Photo)

**PLANTER WALL
SECTION**
SCALE 3/8" = 1' – 0"

A drawing of one of the planter
walls, designed to enhance the
aesthetics of the roadway and
fit into the adjoining neighborhoods. (TxDOT Photo)

A view of the sunken and widened reconstruction, looking north. Visible along the top of the median is the finished result of the proposed planter wall. (TxDOT Photo)

A sample of the numerous design elements built into the final reconstruction. (TxDOT Photo)

The High Five Interchange is the first five-level stack interchange ever built in Dallas. Located at the junction of the Lyndon B. Johnson Freeway (I-635) and Central Expressway (US 75), it replaces this antiquated partial cloverleaf exchange constructed in the 1960s. (Texas Transportation Institute)

Each neighborhood along the expressway wanted an architectural motif of its own, such as the one shown here, to designate and represent its character. (TxDOT Photo)

Approaching the High Five Interchange in Dallas. (Kevin Stillman/TxDOT)

As tall as a twelve-story building, the interchange consists of just under 60 lane-miles of new roadway (comparable to the width of the entire Dallas–Fort Worth Metroplex), stretching 3.4 miles east and west and 2.4 miles north and south. The Interchange was constructed by Zachry Construction. (Texas Transportation Institute)

A magnetic nail picker from the 1930s. (TxDOT Photo)

Pulling It All Together

The State of Texas, by its very size and nature, has always been separated into smaller "countries" defined by their geographical diversity. These parcels of territory, marked by settlement, landscape, and intent, may lie adjacent to one another but are traditionally as divided in their cultures as France and Italy, the Netherlands and Spain. The people of these lands may all speak English, but they shape their own versions of it. East Texas, with its Deep Southern accents and cuisine, its wild dogwood trees blooming among the pine forests, its lumber industry and cotton plantations, had previously been as unrelated to the arid plains and the sweeping, empty desert ranges of cattle-based West Texas as the mesquite-scrub, Hispanic-adobe region of South Texas was to the river valleys, limestone mountains, and German architecture of the Central Hill Country, and the salt-tanged, kelp-strewn beaches of the Gulf Coast. But by tying all these "nations" together with a smooth web of roads, Dewitt Greer and his predecessors and colleagues had finally managed to unite them into one strong, unified identity: the Texas we know now. And with this task achieved, the next leap of progress was about to begin under Greer's stewardship—just in time to realize an even more gigantic and long-rooted dream.

A 1960s model of the nail picker.
(TxDOT Photo)

A bold new identifying decal
appeared on vehicles in 1952
as the department stood on the
brink of the largest construction project in modern history.
(TxDOT Photo)

Manpower and elbow grease in 1949. (TxDOT Photo)

Governor O. B. Colquitt breaking ground for the Post Highway in Austin in 1914. (Courtesy of Texas State Library and Archives Commission)

Ribbon Cuttings and Ground Breakings

G round breakings and ribbon cuttings have long been an important community tradition to signal beginnings and endings for important infrastructure projects.

Former Highway Commission chairman Marshall Formby uses a blowtorch to cut the ribbon for the opening of the Devils River Bridge. Looking on is Val Verde County Judge Jim Lindsey, the "little devil" Elizabeth Henderson, Miss Del Rio Beth Ward, H. B. Zachry, Del Rio District Engineer C. N. Parsons, and Hilary Doran from the Del Rio Chamber of Commerce. (Jack Lewis/TxDOT)

Guadalupe River Bridge opening at Cuero in 1938. Dedicatory addresses were made by Robert Lee Bobbitt, chairman of the Highway Commission, and T. H. Webb, assistant highway engineer. (TxDOT Photo)

Frances Ione Webb, daughter of Assistant Highway Engineer T. H. Webb, cut the ribbon opening the Guadalupe River Bridge at Cuero in 1938. (TxDOT Photo)

It was dedication day in Yoakum in 1937 when Gibb Gilchrist addressed the multitudes and then rode through the new underpass in an ox cart. (TxDOT Photo)

Crowds turned out for the opening of the highway through Yoakum in 1937. (TxDOT Photo)

Governor Rick Perry, Commission Chairman Ted Houghton, and Commissioners Bill Meadows and Jeff Austin joined dignitaries, including Executive Director Phil Wilson, for the opening of SH 130 south of Austin. (TxDOT Photo)

Judy Benson (left) and Patsy Mays, from Oak Cliff, set fire to a red ribbon officially opening the final segment of the R. L. Thornton Freeway (I-35E) in Dallas in 1965. Highway Commission Chairman Herbert C. Petry Jr. (left) observes the proceedings warily, and Police Captain P.W. Lawrence drives the first car on the 2.5-mile section. (Marvin Bradshaw/TxDOT)

(Texas Transportation Institute)

The Midcentury

Eisenhower's Dream

We are pushing ahead with a great road program, a road program that will take this nation out of its antiquated shackles of secondary roads all over this country and give us the types of highways that we need for this great mass of motor vehicles.—President Dwight D. Eisenhower, Cadillac Square, Detroit, Michigan, October 29, 1954

This nation doesn't have superhighways because she is rich; she is rich because she had the vision to build such highways.
—State Engineer Dewitt C. Greer

In 1950, exactly halfway through the twentieth century, Gibb Gilchrist co-established a research center for advancements in transportation safety and technologies. After leaving the state engineer position with the Texas Highway Department in 1937, Gilchrist served first as Texas A&M's dean of engineering, then its president, and was now its chancellor. This must have seemed a very fitting next step in his pursuit of excellence in highway and transportation efficiency: to formalize the legislated mandate defined in the 1917 act creating the Texas Highway Department, which stated that Texas A&M and the University of Texas would collaborate with the department in research and testing endeavors. His fellow founders were also people with deep historical commitments to good roads: State Engineer Dewitt C. Greer, fast on his way to becoming the "dean of the nation's highway administrators," and a short time later, Thomas H. MacDonald, who from 1919 to 1953 had held the position of the first chief of the US Bureau of Public Roads under seven different presidents and was already well known across the country as the father of modern highways. "Next to the education of the child, road building is the greatest public responsibility," he had written as his credo.

The Award of Honor was established by the Texas Section–American Society of Civil Engineers to be presented to a limited number of members of the Texas Section "in recognition of service to the Texas Section and outstanding professional achievement in civil engineering." Here, the award is presented to Thomas H. MacDonald (left) and Gibb Gilchrist (center) in 1953. Also pictured are Highway Commissioners D. K. Martin and R. J. Potts. (TxDOT Photo)

With three such illustrious advocates and practitioners of high standards and progress in the "driver's seats," Gilchrist's decision was a landmark. Only one other academic institution in the country had started any similar program—the University of California at Berkeley, in 1948—and the potential for developing new pavement mixtures, safer guardrails, crash cushions, end terminals, barriers, new signage devices designed to protect drivers during a collision, and a host of other innovations and improvements would benefit not only Texan drivers, contractors, and engineers but the rest of the entire planet. The irony of emeritus chief Thomas MacDonald joining the founding team lay precisely in the fact of his lengthy term of service and the brusque nature of its termination; Secretary of Commerce Sinclair Weeks, for political reasons (and at the behest of President Eisenhower), had "retired" him with a phone call summons, demanding he vacate his office by the next day, and replaced him just as that vision was about to be made manifest.

MacDonald had always believed in building roads only when and if they became needed, a policy that apparently did not fit with the futuristically aligned reach and purposes of "broader ribbons across the land." More important, MacDonald had independent nonpartisan control of fund allocations, and his own long-standing set of powerful contacts seemed inimical to Weeks's federal-aid road plans. That to all practical purposes, MacDonald's

A maintenance crew makes routine repairs to US 67 near Greenville. Interesting how safety regulations have changed drastically in the modern era—no hard hats, safety vests, or steel-toed shoes here. (TxDOT Photo)

replacement would be his own assistant, a Texas-trained engineer whom he had raised to the position of coordinator for the Inter-American Highway, was another irony fully realized soon after MacDonald moved to College Station—the place where that assistant had received his engineering degree. Meanwhile, the widowed MacDonald's final act in federal employment was a grand romantic gesture: after learning of his fate, MacDonald returned to his office and told his secretary of many years, Caroline Fuller, "I've just been fired, so we might as well get married." The very next day they left Washington, D.C., for a new home in College Station.

Modeling New Possibilities for the World

People who are today familiar with the US interstate history sometimes hold the misapprehension that, after penetrating Germany at the end of World War II and encountering the Bundesautobahnen (federal roadways) there, General Dwight D. Eisenhower esteemed their wartime advantages mainly as tactical tools of aggression—that the wonderful, smooth autobahns started in 1929 during the days of the Weimar Republic, first opened in 1932 in a section between Cologne and Bonn and finished elsewhere throughout the country by Adolf Hitler in the 1930s, made it not only easier for the Third Reich to invade and "liberate" Vienna and the Sudetenland but also easier for the Allies to vanquish the Nazis in situ. Certainly this has been true throughout the history of human conquests, as it was true for the conquistadors marching along the Mayan roads to vanquish the Aztec Empire of Mexico. But in fact, due to modern changes in warfare, the German autobahns made it harder—which was one reason Eisenhower admired them. The task of effectively bombing a moving convoy spread out over many miles of road surfaces was much more demanding and trickier to accomplish than taking out a munitions supply train on a track or a mass of infantry troops. Even after bombs cratered these excellent roads, military vehicles could still move over them with comparative ease. But whether the autobahns aided national defense or made cutting through the countryside a swifter proposition for the invaders, they inspired Eisenhower to continue the vision he had nurtured ever since the 1919 motor convoy crossed the continental United States and turn it into a reality.

The borders of Texas enclose a landmass that is nearly twice the size of the entire country of Germany. Its square mileage totals more than that of Greece, Italy, and the United Kingdom combined. It is larger than Sweden, Spain, France, or the Ukraine and only slightly smaller than Turkey. Its well-tended roads traverse every part of the state—even the most remote, the most sparsely populated. And Texas, of course, is only one out of fifty states. To conceive of a road system that would include, navigate, and knit together all lower forty-eight states *with a completely uniform, integrated design* was an ambition not only of monumental proportions but of true unifying potential—a physical fulfillment of our nation's very name. And it was, until 2011, the largest single public works construction enterprise ever completed by human beings in the history of the world.

Only China's recently extended expressway network supersedes it in length, scale, and scope. In 2011, China added 6,835 miles of expressway to its national system. Now, at 52,800 miles, China's intercity freeway system is longer than the 46,720 miles in the US interstate system. Before China began procedures to upgrade its road systems, however, no other achievement could compare to the sheer size of the US interstate project—not the pyramids, not the buried cities of the Yucatán, the aqueducts of Rome, the Panama or Suez Canals—not even China's Great Wall. And China's construction has been rapid indeed—sometimes alarmingly so, in terms of safety and solidity. Some of its expressways seem to lead nowhere; visitors have commented on the emptiness of the more provincial roads, the lack of cars traveling in either direction. In regard to the interstate, or "interregionals," as Dewitt Greer first called them (a term dating from early planning sessions), even the most remote parts of Texas are striped by heavily trafficked

Lancaster Hill in Crockett County carried traffic on US 27;
after completion of I-10 through West Texas, the road was
renamed US 290. (TxDOT Photo)

A family traveling from Georgia
stops for photos on El Camino
Real, 1955. (TxDOT Photo)

*I-10 west of Kerrville.
(TxDOT Photo)*

connectors: four major east-west interstate arteries cross the state—10, 20, 30 (paralleling US 67 most of the way), and 40—as well as the all-important north-south route, I-35, and a number of shorter sections, amounting to a total of over 3,000 miles. Included in the shorter sections are the loops around Fort Worth, Houston, San Antonio, and Dallas, in addition to four Texas highway routes that have been upgraded to national standards: I-37, I-45, I-27, and I-44.

In 1954, when President Eisenhower signed the Federal-Aid Highway Act that dedicated an initial $25 million amount annually for highway construction, the money was meant to cover two years' worth of road-building efforts. Immediately, Texas began construction. Then in 1956, when legislation finalized the federal interstate highway system plan and created the annual $175 million Highway Trust Fund to pay for it, each state became

I-45 near Huntsville be-
tween Dallas and Houston.
(Hal Stegman, TxDOT)

responsible for meeting 10 percent of the outstanding costs—quite a fiscal departure from the previous "matching funds" precedent in the federal/ state partnership. Eisenhower encouraged the states' use of bonds to promote their shares of the funding. But as Dewitt Greer told the president during a meeting in Washington, Texas would meet its portion of interstate debt through its usual pay-as-you-go policy. When Eisenhower expressed his concern about this, in case it held up the progress of the construction or obviated it altogether, Greer told him, "That's how we do it in Texas. It may take a little longer, but we'll succeed," and assured him that the results would be perfectly satisfactory.

Dewitt Greer himself truly came into his own during this early period of interstate development in the 1950s. Already he had served longer than any other state engineer since the department's founding; in 1958, his tenure would have lasted twice the duration of the longest continuing stint, that of his mentor, Gibb Gilchrist. The ongoing application of Gilchrist's tenets had by now molded the department into an almost unassailable structure of efficiency, progressive energy, thrift, and uprightness, and Greer personified and directed those qualities. If he dominated his domain with an appearance of absolute power (as, according to numerous people, he did), it was a benign and positive power—a verity to which many colleagues and peers attested over and over. "Greer was the most effective public administrator I have ever known," Governor Dolph Briscoe said of him later. "He was a person of great character and integrity," said Bartell Zachry, chairman of Zachry Construction, one of the Texas Highway Department's longtime contractor partners in road building. "He wouldn't tolerate the slightest infraction that might reflect adversely on the highway department."

D. C. Greer. (TxDOT Photo)

Greer's determination to run a clean agency and personal life also precluded the tiniest act of obligation. Indeed, he grew famous for never accepting so much as a cup of coffee from any contractor anywhere—an ironclad rule made even more noteworthy by the fact that such an offer did not legally become an ethics issue until the Bribery and Corrupt Influence Statute, defining gifts proffered to a public servant in return for favors (beverages included), was incorporated into the Texas Penal Code in 1973. In this way he could claim sole responsibility and sovereignty for his decisions, except when they were preempted by federal regulations and mandates—which happened a short time into the interstate construction process.

Doug Pitcock, the cofounder, chief, current board chairman of Williams Brothers Construction Company, Inc., and one of the most influential transportation experts in America, tells a story of his efforts in trying to build up a young business at a time when all highway job bids were submitted in person. "Back in those days we had a highway letting once a month, and before computers, half the state came to Austin to either bid work or quote materials and subcontract prices to contractors that were bidding work. [On September 11, 1961, Hurricane Carla, the most intense US tropical cyclone landfall ever recorded on the Hurricane Severity Index, was scheduled to make landfall from the Gulf of Mexico.] Every time there's a hurricane down here, people run and they should. . . . You run from the water, but you don't run from the wind, and the people in Beaumont and Port Arthur and Orange, Lake Jackson, Galveston, they all go to Austin. So that hurricane was coming in about three days before the highway letting, and the highways were bumper to bumper . . . and when the highway letting's in town, they

Recently inducted into the American Road Builders Hall of Fame, founder and board chairman of Houston's Williams Brothers Doug Pitcock surveying construction near Houston. (Courtesy of Doug Pitcock)

Governor Price Daniel and State Highway Engineer D. C. Greer put the finishing touches on one of the signs to go up on a traffic safety project. (TxDOT Photo)

take every hotel room in Austin. Well, these people running from the hurricane were taking all of our hotel rooms." From his office in Houston, Pitcock telephoned the Association of General Contractors (AGC) to ask if they could convince the Highway Department to cancel the letting, as very few would be able to attend, much less participate. A few hours later, the head of the AGC phoned him back to say his request had been refused. "I guess it was a Saturday afternoon," Pitcock continues. "I picked up the phone and I called Information in Liberty, Texas, and I said, 'I want the phone number of Governor Price Daniel.' Well, they gave it to me. So I dialed the phone and this guy answered and I said I'd like to speak to Governor Daniel. He said, 'Well, this is he. What can I do for you?' So we started talking and I told him . . . and he started laughing. He said, 'Well, you're right, Highway 90, and they're all headed for Austin right now.' But he said, 'Mr. Pitcock, you know I can't tell Dewitt Greer what to do.'"

As another unnamed Texas politician remarked during the same period, "Ask me to do anything but try to influence the Texas Highway Department. Nobody bothers Dewitt Greer." Yet the interstate requirements would soon bring Greer's authority under stress and stir criticism for his compliances—particularly when it came to building overpasses in forsaken spots and roads across empty vistas.

US 81, now I-35, entering San Antonio in 1956. (TxDOT Photo)

The Interstate

In 1956, when Eisenhower's cherished dream was on the brink of fulfillment, four key Texans contributed to its facilitation. Foremost among them of course was Dwight D. Eisenhower himself—a Denison native transplanted to Kansas during his boyhood years. Immediately under him was Dallas-born, Texas A&M class of 1929 graduate Francis C. Turner, whom Eisenhower eventually, in 1957, placed in charge of interstate project negotiations after initially selecting him in 1954 to be executive secretary to the President's Advisory Committee on the National Highway Program. According to his *New York Times* obituary, "Francis C. Turner, often called the chief engineer of the Interstate System of highways that redrew the map of America . . . was the only person to rise through the ranks to head the Federal Highway Administration; he worked forty-three years there and at its predecessor, the Bureau of Roads." At the time the Federal-Aid Highway

Act of 1956 went before Congress, Sam Rayburn, the forceful political influence from Bonham, was Speaker of the House. The Senate majority leader (or, as Robert Caro so aptly named him, the "Master of the Senate") was the former teenaged Texas Highway Department employee from Johnson City, Lyndon Baines Johnson. Eisenhower signed the act into law on June 29. The wording of the legislation required that the entire federal system be completed by 1972—a tall order for sprawling Texas, despite the way it had already loped ahead of most states in its construction endeavors.

Even the lesser-known contributors often held Texas ties. The sign that would mark every interstate highway across America with its distinctive shield shape; red, white, and blue color scheme; and specific designation numbering—the only trademarked highway marker in use in the United States—was the product of a competition staged between transportation departments throughout the country. The winning design came from Richard Oliver, a Texas Highway Department employee in the Maintenance Division.

Letter from D. C. Greer informing Richard Oliver that his design had been selected for the interstate road shield. (TxDOT Photo)

Richard Oliver and his winning interstate highway shield design. (Gay Shackelford/TxDOT)

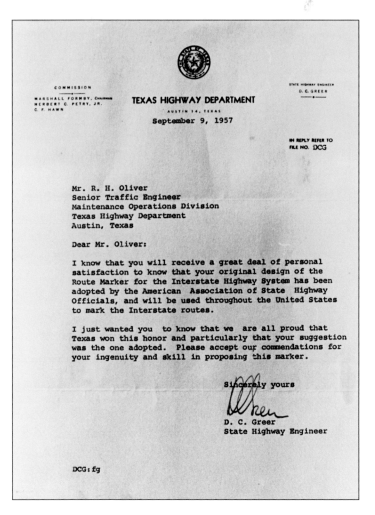

COMMISSION
MARSHALL FORMBY, Chairman
HERBERT C. PETRY, JR.
C. F. HAWN

STATE HIGHWAY ENGINEER
D. C. GREER

TEXAS HIGHWAY DEPARTMENT
AUSTIN 14, TEXAS
September 9, 1957

IN REPLY REFER TO
FILE NO. DCG

Mr. R. H. Oliver
Senior Traffic Engineer
Maintenance Operations Division
Texas Highway Department
Austin, Texas

Dear Mr. Oliver:

I know that you will receive a great deal of personal satisfaction to know that your original design of the Route Marker for the Interstate Highway System has been adopted by the American Association of State Highway Officials, and will be used throughout the United States to mark the Interstate routes.

I just wanted you to know that we are all proud that Texas won this honor and particularly that your suggestion was the one adopted. Please accept our commendations for your ingenuity and skill in proposing this marker.

Sincerely yours

D. C. Greer
State Highway Engineer

DCG:fg

And Frank Turner himself, the Texan credited as the chief architect who eventually earn the title "Father of the Interstate," supplied a simple but apt definition of the network and its functions: "A system of highway pieces all joined together so that you could get from anywhere to everywhere . . . a massive concept, all right." Meanwhile, Dewitt Greer ramped up plans to implement construction, revealing their details through "recommendations" contained in this letter written on May 10, 1956, and addressed to the highway commissioners then in office:

Honorable E. H. Thornton, Jr., Chairman
Honorable Marshall Formby, Member
Honorable Herbert C. Petry, Jr., Member

Gentlemen:

Congressional approval is now assured on a thirteen year Interstate Highway System program, involving Federal Funds in the amount of $25 Billion and State Funds in the amount of $2 Billion. In round figures Texas' share in this thirteen year program will be the $754 Million in Federal Funds matched by $83 Million in State Funds, to constitute an over-all thirteen year program of $837 Million on the Interstate System. This Bill establishes Congressional intent over the thirteen year period but only allocates the money for two years of operation.

The Act provides that funds for the first two years shall be apportioned and made available for expenditure immediately upon the enactment of the Bill. This means that, insofar as Texas is concerned, immediately upon the passage of the Act we have available to us on this particular system Federal Funds in the amount of $102 Million. This requires $11 Million of State matching funds, making a total program amount of $113 Million. It is our recommendation that we proceed at once to form a work program, by projects, covering this amount of money. We recommend the following policy procedure.

Since Congress has indicated by its action that they expect to reimburse the States the amount of money that has been spent on the Interstate System on such sections as meet the Interstate standards, and if this work has been placed under construction prior to June 30, 1957, it is felt that our interests would be best served by making the No. 1 priority in this program the bringing up to standard those portions of the Interstate System that we have partially completed to Interstate Expressway standards, as rapidly as possible.

It is recommended that, in fairness to those counties and cities that have heretofore secured adequate controlled access right of way, as requested by us on the Interstate System, we should include projects in the program to take advantage of this right of way, regardless of their comparison in merit with other portions of the system where right of way has not heretofore been furnished.

After we shall have withdrawn the necessary funds to accomplish the objectives covered in the two preceding paragraphs, it is recommend that we take the remaining total and allocate it by Districts, based upon the following criteria.

25% of the total shall be separated out and classified as available for work within urban areas on the Interstate System. These funds shall be distributed on the basis of the ratio that urban population of the District on the Interstate System bears to the urban population of the state as a whole on the Interstate System. The remaining 75% of the total will be distributed based upon the ratio of the rural vehicle miles of travel on the Interstate System in the District as related to the total state-wide rural vehicle miles of travel on the Interstate System.

The funds apportioned as above indicated shall be used for the purchase and fencing of right of way, for the removal, relocation or adjustment of utilities where such utilities are owned and operated by a governmental agency, and for the total construction of the highway facility.

It is requested that you carefully consider these recommendations and give to us your policy decision, in order that we may proceed with this development, bringing the projects back to the Highway Commission for approval in order that engineering work may proceed.

Sincerely yours
D. C. Greer
State Highway Engineer

If any serious dissent fomented among the commissioners regarding these suggestions, no ghostly rumor of it stands out now. One of their number, Herbert C. Petry Jr., who was presently appointed Highway Commission chair, later noted with some jocularity, "We on the Commission, all being U.T graduates, have learned you can work with a Texas A&M graduate—if you *must*." And as Price Daniel remarked to Doug Pitcock, not even the governor could dare tell Dewitt Greer what to do.

But reading between the lines of Greer's letter, the pleasurable anticipation inherent in the impending challenge still to this day sounds obvious. The organizational thoroughness with which he had already formulated what lay ahead, the equity with which he approached the issues of districts and counties, the opportunities for doing a good job, the rewards of a job well done to look forward to: all can be inferred from his full-steam-ahead prose. His instructions to his district engineers sounded just as clear: Make it wide enough and smooth enough, with broad interchanges, to assist

Highway Commissioner H. C. Petry cutting the "ribbon of girls" at the opening ceremonies of I-35 in Austin, March 29, 1962. (TxDOT Photo)

truckers transporting goods and services. Design expansive shoulders so that drivers can pull off when necessary. Hold contractors to tight specifications and high standards. Purchase and obtain enough right-of-way to accommodate the widening of roadbeds and the increase in lane numbers in the future. As the Associated Press Bureau journalist Garth Jones so presciently observed in 1957, "A long, smooth stretch of Texas highway does the same thing to Dewitt Greer that a sunset does to other persons."

Controversy

Just as Central Expressway construction, a few years before, had required the razing of large sections of historic properties, demolishing retail businesses and homes, and splitting the socioeconomy and geography of Dallas like a cloven apple, so the new interstates rampaged through previously respected and even sacredly esteemed areas of Houston, San Antonio, Austin, and elsewhere, wiping out movie theaters, hotels, churches, parks, blocks of crumbling houses, and modest neighborhoods, crushing smaller communities dependent on the traffic from older roads for their livelihoods, closing gas stations and cafés located beside the old highways, and sending many people, at least those who could afford it, to live in the newly developing suburbs. For instance, the town of McClean, once a thriving stop on Route 66 filled with 1,500 people and fifty-nine businesses, in 1984 became a casualty bypassed by I-40, when the population dropped to 830 and most of the businesses were shuttered forever. Urban sprawl and new suburbs,

This is the main entrance to McLean in 1938 on Route 66, much earlier than the interstate era. The town was eventually bypassed by I-40. (TxDOT Photo)

highways diverted away from small centers, the traffic congestion in cities where local residents still used the old paths that the interstate had now replaced every day for work commutes and general conveyance around town (such as I-35 in Austin), the strip malls and office parks along frontage roads that changed the character of old neighborhoods and zonings—such as NorthPark Center in Dallas, the first shopping center ever featured in *Vogue* magazine, that opened in 1965 at the intersection of I-75 (North Central Expressway) and Loop 12 (Northwest Highway)—these changes would forever alter the culture and economics of life in Texas.

The road designers justified this apparent ruthlessness with the argument that much of what fell under demolition in the cities was urban blight and that city life in general would enjoy great improvements of all kinds with the new systems in place—especially the all-important movement of goods and services, which entailed wide, even roads on which trucks could

Construction of I-35 (Stemmons Freeway) through Oak Cliff plowed through numerous low- to middle-class neighborhoods in the late 1950s. Here the freeway has reached Marsalis Avenue (the Dallas Zoo is the upper left portion of the intersection). The interstate was eventually completed, traveling south to Laredo. (TxDOT Photo)

This image shows scope of interstate construction through Austin just south of the Colorado River. The Riverside Drive overpass is seen at the bottom of the hill in the background. (Courtesy of Holt Cat)

By 1974, I-35 was completed through central Austin; this view looks north with the Riverside Drive overpass in the center of the photo. (John Suhrstedt/ TxDOT)

turn, as well as many slots of access for more stationary citizenry. Because of Dewitt Greer's philosophy, that roads are made for local people, not just drivers and truckers passing through, the Texas interstates would have more points of entry than most in other states, in addition to more frontage roads. Nonetheless, major conflict broke out over the routes and continued to explode through the 1960s and 1970s in San Antonio and other municipalities. Some people went so far as to take legal steps to prevent the progress of interstate routes, such as a group of citizens in Amarillo who filed

Old Route 66 through Vega, Texas. (J. Griffis Smith/TxDOT)

a class-action injunction against State Engineer D. C. Greer, the members of the Texas Highway Commission, and the City of Amarillo, its mayor, city commission, and city manager, in a futile effort to stop I-40 from slicing through their central Panhandle town. The blowback Dewitt Greer received during the decades-long process of interstate construction, however, focused even more on his bowing to the restrictions and prerequisites of the federal guidelines. These compelled him (often against his own judgment) to build what his critics called "overpasses for jackrabbits" in some isolated tract of West Texas—even though he personally saw no need for full access control on highways in sparsely populated areas.

All in all, the interstates can fairly be said to have contributed a new spectrum of experience and opportunity to Texans. As former Houston mayor Bob Lanier later opined, "We'd be far less wealthy, a far lesser quality of life, had the Interstate system not been built." But Walter J. Humann, chairman and CEO of the WJH Corporation in Dallas and a crucial provider of good transportation solutions, summed it up even more justly: "Different day, different climate, different criteria for what was good and bad, and they just got the dirt flying and got on with it."

Sunrise along I-10 in Culberson County. (Kevin Stillman/TxDOT)

Expressway and commercial development on US 81 looking north from R. L. Potts interchange in Waco in 1956 (now I-35.) (TxDOT Photo)

The first section of interstate highway to be let in Texas, in 1956, lay outside Corsicana in Navarro County on what became I-45. The last, that of I-27 between Lubbock and Amarillo, reached completion thirty-six years later. When Dewitt Greer stepped down from the state engineer position in 1968, the program was already 60 percent completed, but there were still about 1,400 miles remaining to be built. For instance, the nearly 15-mile stretch of I-44 connecting Wichita Falls with Oklahoma was constructed and/or expanded entirely in the 1980s; I-10 from San Antonio to El Paso did not open in a finished form until the 1990s. By that time, the federal 1972 deadline was already ancient history. But then, the other states' tasks seldom compared with a fraction of what Texas had to accomplish.

Explosives, bulldozers, and trucks slice through the Central Texas Hill Country during construction of I-10 near the town of Junction. (Jack Lewis/TxDOT)

District Engineers

The district engineers—frequently the unsung heroes of the department, although always its public faces in the twenty-five regional district locations—acted to smooth the paths toward interstate right-of-way acquisitions, oversaw construction, and worked with the small-package, local section contractors that first Gibb Gilchrist and now Dewitt Greer encouraged.

Throughout departmental history, the importance of the district engineers, with their range of skills, can be neither overstated nor overpraised. Often their apprenticeship encompassed long periods; always it proved arduous. For this reason, the multiple decades of many highway department careers—especially those begun in the postwar years—could be attributed not only to agency loyalty, job dedication, and the joy of problem solving but to a sense of pride at having achieved such a solid pinnacle of expertise. "To get to be a resident engineer in the 1950s, 1960s, and 1970s, it took you eighteen years—to learn how to maintain, how to construct, how to design, how to buy right-of-way, how to deal with people, how to hold public hearings," explained Marcus Yancey, thirty-six-year Texas Highway Department veteran engineer and former deputy executive director. "It's not taught in schools. It's not taught anywhere else. You have to do it. And out of that . . . you can shake the milk bottle, but the cream always comes to the top. And we would watch people rise in the department, and they achieved it by merit. You can see on their records, nationally and state-wide, what they've accomplished."

Other Irons in the Fire

The contributions of Texas to the national interstate process were multiple. From the first state Materials and Tests Division, established by Gibb Gilchrist and solidified by Dewitt Greer, which became a model for other states throughout the nation, to the first multidiscipline design team in the United States, and even the first public information officer representing a transportation department, Hilton Hagan, Texas demonstrated a structure and approach to development problems that served the country on a broader scale. But it also directed these resources toward its own non-federal concerns. From 1955 through 1957, the department contributed know-how and supervision to at least one major arterial project that predated the interstate routes but that ultimately would be absorbed into I-30 (now running from I-20 west of Fort Worth to Little Rock, Arkansas). The very first toll road in Texas was designed to link two of the most important cities in the state. Before its construction, an obstacle course of eighty-two stoplights slowed traffic between Dallas and Fort Worth to a 30-mile stutter. Drivers who began their journeys on US Highway 80 in Dallas and continued on to Division Street in Arlington could expect to endure a long, wearisome crawl before finally reaching the Fort Worth city limits, sometimes hours later. Residents of the small settlements scattered east and west in between had little hope or expectation of connecting easily with their neighbors.

Hilton Hagan. (TxDOT Photo)

This field test in Fort Bend County was conducted in 1935. It was a precursor to the stringent testing standards implemented by Highway Department leaders during the ensuing years. (TxDOT Photo)

DFW Turnpike entering Dallas, 1957. (HNTB)

Then in 1953, the state legislature created the Texas Turnpike Authority. The next year the new authority "borrowed" head of the Texas Highway Department Road Design Division J. C. Dingwall, who had originally come to work for the department in 1928 under the aegis of Gibb Gilchrist and had remained there until the war years, which he spent building airfields internationally with the Army Corps of Engineers. The authority named Dingwall its new engineer-manager and charged him with the task of creating the Dallas–Fort Worth Turnpike—a project that had been rejected by Dewitt Greer in 1943 as being too expensive for the department to attempt at the time. In 1955 this new agency raised $58.5 million (equal to roughly $515,478,862 in 2014) in bond funding to build Texas' first toll road, a throughway composed of three lanes in either direction that would reduce the time of the trip between the two metropolises to half an hour.

The entire package, including the design, the right-of-way acquisitions, and the construction, took only three years to complete. It opened in August 1957, the bonds to be paid back through income derived from the tolls—a conclusion that took twenty years, seventeen years ahead of schedule, at which point the turnpike was turned over to the Texas Transportation Department (formerly the Texas Highway Department) as the free public-use Interstate 30, and the toll booths were removed within the same week. Immediately on completion of construction in 1958, the communities near the route, particularly Arlington and Grand Prairie, began

DFW Turnpike entering Fort Worth, 1950s. (HNTB)

DFW Turnpike Service Plaza, 1950s. (HNTB)

to flourish. Arlington's population increased dramatically from thirty-four thousand the first year to forty-nine thousand three years later, and in 1961, the theme park Six Flags over Texas opened its gates there to usher in a new era of tourism and mounting affluence for the area. In 1958 J. C. Dingwall returned to the department and was given the job of assistant state highway engineer—the equivalent of deputy executive director today.

Another project, entirely instigated and launched by the department, began to rise over the water at Corpus Christi between 1956 and 1959. A new Harbor Bridge replaced the old drawbridge; as the second-tallest bridge in Texas when completed, its road deck soared 138 feet above the waves. The design was the product of the Texas Highway Department Bridge Division's Vigo Miller, with the firm of HNTB acting as consulting engineers. The bridge's riveted, cantilevered truss construction and tied-arch center span, with precast, prestressed, post-tensioned concrete beams in the approach spans and the revolutionary use of neoprene pads as bearing plates, made it not only an architectural icon for the time and the city but an example of the Texas Highway Department's pioneering of new problem-solving materials.

When Six Flags over Texas was first established in Arlington in 1961, there was sparse development. Today, Arlington is a bustling city and home to the Texas Rangers, Dallas Cowboys, and numerous tourist destinations. The DFW Turnpike can be seen in the background at the top. (TxDOT Photo)

Harbor Bridge, or Corpus Christi High Bridge, on US 181 crosses narrows between Nueces and Corpus Christi Bays. The bridge over the shipping channel is 235 feet high with a 640-foot main span. (Jack Lewis/TxDOT)

Afterward, neoprene bearing pads became the standard specification incorporated by the American Association of State Highway and Transportation Officials (AASHTO; the American Association of State Highway Officials [AASHO] added "Transportation" to its name in 1973) for nationwide use to mitigate the effects of seismic events, concrete creep, and traffic vibrations.

The Ongoing Process

During the 1950s, 1960s, and 1970s, as work on the interstate system proceeded, a number of other innovations emerged from the collaborations between the Texas Transportation Institute, the Center for Transportation Research (established at the University of Texas in 1963), and the State of Texas that led the nation and the world in safety, road materials, and technological advances for building and maintaining highways. Among these improvements were prestressed concrete development, slip form pavement sections, asphalt binders, grooved pavement, breakaway signs and collapsible or breakaway rural mailbox supports, better signage, and, eventually, computerized engineering. An array of crash cushions have evolved through the years to better ensure the survival of a car's occupants in collisions with barriers, walls, and guardrails by absorbing the force of the impact. A partial list of other safety devices includes these products:

Single slope barriers
Low-profile concrete barriers
Steel-reinforced safety poles
Barrel crash cushions
Roadway illumination systems
Pavement skid resistance
Energy-absorbing bridge rails
Safe sloping culvert grates
Texas T-I bridge rail system
Texas concrete median barriers
Aluminum bridge rail systems
Concrete barrier end treatments
Pavement edgers for vehicle stability
E-T 2000 guardrail end terminal

Some of the other long-term major contributors to the interstate system, as well as scores more projects in partnership with the Texas Highway Department, are the contractors who carried out the designs and the labor—large companies such as Williams Brothers, J. H. Strain, the Zachry Construction Corporation, Hunter Industries, and HTNB, as well as the large numbers of smaller, local contracting firms. The thrift with which Dewitt Greer directed the department's operations is demonstrated by the 1963 statistic, which summed up the average cost per mile of an interstate

The Texas Transportation Institute has invented and improved a number of life-saving innovations for road construction, including the breakaway sign. (Texas Transportation Institute)

highway as $1 million throughout the United States—except in Texas, where the cost was $610,000 per mile.

By the time Greer retired from the department in 1967, he had supervised the spending of nearly $5 billion, unaccompanied by the slightest trace, breath, or even suspicion of scandal or malfeasance, with a monumental result for the money invested—Texas' part in the largest construction program in the history of the world at that time. The rate of increased traffic on the interstates of Texas, where 30 to 40 percent of usage is accounted for by trucks moving goods and services and the rest is made up by private-citizen drivers, can be traced through vehicle registration figures, starting with the first postwar boom in car acquisitions:

1945	1.6 million
1950	3.0 million
1955	3.9 million
1960	4.5 million
1970	6.4 million
1980	8.4 million
1990	12.4 million
2003	14.0 million
2012	22,780,321

Texas is now the second most populous state in the nation. Its interstate miles comprise more than those of any other state. Despite the wear and tear on the highways caused by truck traffic (an eighteen-wheeler does ninety-six hundred times the damage on a flexible pavement that a luxury car does, according to an AASHTO study), Texas has a low registration charge on trucks—as low as $110 per year and up to $840 per year for the highest legal weight (8,000 pounds while empty)—compared to California, where an eighteen-wheeler's registration costs $2,064 per year. The financial implications for the continuing health of Texas interstate highway maintenance, as well as that of other, "lesser" roads, are obvious.

Z. H. "Andy" Anderson when he retired in 1958. He recalls living on $10 for half a month, the Depression years, and "lots of ups and downs" with the Highway Department. His philosophy is simple: "That goes with life any way you travel." (TxDOT Photo)

My Summer on I-45

Larry Norwood

In 1959 I needed a job between my graduation from Corsicana High and my enrollment in Baylor University.

I knew that construction was underway on I-45 about ten miles north of Corsicana near Alma, an almost-town.

I drove my 1948 Ford up to the construction site and asked to speak to the construction supervisor. I was directed to Buck Wooley, who looked for all the world like his name. He was tall and rangy, with long red sideburns and a battered straw cowboy hat covering rusty-red hair. I introduced myself and told him I was looking for a job for the summer. He asked me whether I had been working. I told him I'd worked for a commercial beekeeper for the past six years and that I figured a construction job would pay a bit more money, which would help me with my Baylor expenses. He whistled to a carpenter and said, "Tom, this here boy's name is Larry. He says he wants a job out here. If he wants to work, find something for him to do. If he don't, show him right there's the road." I've kept those words in my memory for the past fifty-five years, as a reminder of what it means to take on a job: if you want to get paid, do the job. If you don't, there's the road.

The job was building culverts in creeks running under the interstate. A good part of my job was dragging 20-foot reinforcing bars out of a stack down to the hole where the culvert was to be built. The carpenters built wooden frames about ten or 12 feet high. One wooden wall would be lowered into the hole by the dragline operator, who was called "Easy Money" because he refused to do any work other than operate the dragline. After the wall was set in place, a grid of rebars was tied together with wires twisted by a handheld device called a pigtail, a wooden handle with a rotating hook. Once the second side of the wall was in place, with the grid enclosed, everything was squared up and Easy Money was called in to fill between the walls with buckets of concrete. This process was followed until there was a concrete barrel forming the culvert.

The rebars had to be placed an exact distance apart to meet specifications. Occasionally a young inspector would come out and check the spacing. If it wasn't correct, he'd make us untie the grid and start over. Wooley told me a joke: A woman looked out her screen door and saw an inspector and a construction foreman coming toward her house. She went scurrying to get her teenage daughter out of sight. Her husband said, "You don't have to worry about our daughter. Those two guys are too busy trying to screw each other to be thinking about her."

A couple of my friends joined the gang, and I picked them up early enough each morning for us to get to the construction site. When it was raining, we still had to go to the work site to see whether the rain would lift. We didn't get paid for rain days.

There was no shade where we were working, and the temperature often was above 100 degrees. For our twenty-minute lunch we would crawl under a flatbed truck for the shade and eat our lunches. My mother often fixed me a fried ham sandwich. And a thermos of iced tea.

The workers on the construction crew were itinerant. Some, like Tom the carpenter and his brother, worked steadily for the construction company, going from project to project. Others would work a short while, draw a paycheck or two, and disappear.

Buck Wooley had a mean streak. He once got into a fistfight with his superintendent, a guy from the company office named Meadows. Another time he dropped a dead rattlesnake down into the hole onto the neck of a small black shoveler called Cat Man. Cat Man tore out of the bottom of the hole and ran thirty yards before he stopped. He came back and said, "Oh, Mr. Wooley. Please don't ever do that again." Wooley laughed.

Among the workers were a father and son pair named Poole, who liked to drink beer at night and by day to try to get me to get them dates with Corsicana High girls.

There was an older laborer named Phillips who at least twice a day would say, "It's a damn good thing I was born a boy, or else there would have been a whore in the family for sure!"

That summer of 1959 was a good summer. For the first time I found myself working as hard and as long as the cussing and complaining men who were doing that work in order to make a living. I learned to hold my own, and I learned that if you don't want to do the work, right there's the road.

Highway Crime

Throughout recorded history, roads have provided opportunities for criminals. A person on the move makes easy quarry—naked to the elements, for all practical purposes, and therefore ill defended, the approaching target of a lurking predator who can then make a quick getaway after the act. From legends to biblical parable, from Theseus on his way to Athens to the mugging victim aided by the Good Samaritan, a traveler on foot often meant a traveler headed toward a violent destiny. Vehicles did not necessarily confer invulnerability: look at the charioteer King Laius and his assassin son Oedipus; look at Dick Turpin the highwayman; look at Black Bart (Charles Bowles) and how many Wells Fargo stagecoaches he robbed (twenty-eight)—without once firing a gun. Not to mention the numerous bandits who lay in wait around the trails and tracks crossing Texas and elsewhere to pounce on everything from buckboards to covered wagons to full-scale railroad trains. But since the development of better roads, thieves, crooks, and killers have also had the advantage of movement themselves— swift arrivals, speedy felonies, lightning departures—as well as more efficient smuggling operations. The ways in which the Texas Highway Department unwittingly facilitated a greater number of citizens' activities than perhaps it ever intended are demonstrated in the following stories.

Border Bootleggers

In the 1920s, 1930s, and 1940s, *tequileros*, horse-mounted bootleggers bringing mescal and tequila into Texas, used the river crossings and trails previously established by gunrunners through canebrakes and brush country to sneak their wares from Mexico into Texas. But it was the "long-range bootleggers" on the Texas side, buying the *tequileros'* products for out-of-state distribution, who presented a greater danger to lawmen, both during Prohibition and after its rescission. Because these Anglo smugglers (often bankrolled by gangsters from Chicago and northern states) drove automobiles on the many roads and highways threading South Texas, they could obey a Texas Ranger's or DPS officer's order to stop their vehicles while keeping their hands hidden from view. Then they would pull a gun, thus increasing the ever-mounting death count of peacekeepers murdered in the line of duty. They also had a clearer shot on anyone following them than the *tequileros* did, due to the well-kept roadbeds of the early 1930s, courtesy of the Texas Highway Department under State Engineer Gibb Gilchrist.

The long-range bootleggers and whiskey outlaws did not cease their major-market trade runs when Prohibition ended. Instead, they employed an array of transport, from coffee trucks, undertakers' wagons, and hay loads, to school buses (often still used today by drug smugglers camouflaging their contraband through South Texas) as disguise for large-shipment liquor crates bound for the north. County sheriffs and other authorities often set up roadblocks at

Border Patrol agents pictured with a haul from busted border bootleggers near Shafter, Texas, 1928. (DeGolyer Library, Southern Methodist University, Lawrence T. Jones III Texas Photographs)

state-paved intersections, well out of populated areas so that citizens would not get caught in the line of fire, and deployed machine guns in highway battles. East Texas also provided the route between "wet" Louisiana and "dry" Oklahoma during those years of the 1940s and 1950s, with trucks sometimes carrying as much as $9,000 worth of booze, immediately confiscated by the sheriff. In 1943, one dry-county sheriff seized a total of $15,656 worth of illegal liquor and twenty-seven automobiles (some very expensive) on their way to service military bases.

On The Lam

Bonnie Parker and Clyde Barrow, the outlaws who robbed and killed their way into infamy (at least nine peace officers, plus several civilians, including a young father), spent nearly two years on the run, making grand use of the Texas highway system alongside their confederates, starting in 1932.

Although the Barrow Gang drifted to other states in the course of their migrations, ranging as far north as Illinois and Minnesota, their circumlocutions always brought them home to their native state. Their seemingly impulsive progress can be charted on the official 1926 Texas Highway Department map, showing routes that often switched from newly paved state throughways to dusty back roads. The fact that there appeared to be no logic or planning to their rovings made them harder to arrest. Finally, the former Texas Ranger Frank Hamer (who had now joined the Texas Highway Department's Motor Patrol) carefully studied the routes and detected a pattern that revealed Clyde Barrow's thought process. It seemed the gang clung to family reunions. They were also aware of the rules that prevented one law officer from chasing them over a state line into another jurisdiction. So they kept running in wide circles that would

Bonnie Parker and Clyde Barrow. (FBI)

Bonnie and Clyde wanted poster. (FBI)

bring them back for regular visits with kinfolk. This insight helped Hamer track the pair, get permission to lead a six-man posse across the line to northern Louisiana, stage an ambush through a gang member's father, and kill the couple there, in a storm of gunfire, right in the middle of the road—appropriately, while sitting in their car.

Highway Hunters

Since the introduction of the Interstate system in the 1950s, serial killers have used the smooth roads of Texas and elsewhere to their great advantage—so much so that in 2009, the FBI publicly launched the Highway Serial Killings Initiative, a project to trace and interconnect murders from state to state and pool data that would lead to the apprehension of the murderers at work. Several interstates in Texas have been particularly useful to these predators. According to the FBI's website, "In 2004, an analyst from the Oklahoma Bureau of Investigation [a female Supervisory Intelligence Analyst named Terri Turner] detected a crime pattern: the bodies of murdered women were being dumped along the Interstate 40 corridor in Oklahoma, Texas, Arkansas, and Mississippi. The analyst and a police colleague from the Grapevine, Texas Police Department referred these cases to our Violent Criminal Apprehension Program, or ViCAP, where our analysts looked at other records in our database to see if there were similar patterns of highway killings elsewhere. Turns out there were." According to Mike Harrigan, head of the FBI's ViCAP, "The Interstate Highway System is a plus for these highway serial killers to act out their fantasies." Much of the time, says the FBI, the offenders are long-haul truckers who have joined that employment precisely to satisfy their murderous needs and intentions.

In the late 1970s, around twenty women in Texas who were either stalled beside the highway with car trouble or hitchhiking became victims in a streak of murders known as "The I-35 Killings." Several of these were later credited to the serial killer Henry Lee Lucas, who was arrested in Texas in 1982 after a long-running spree of highway murders—as well as a few stationary ones. Robert Ben Rhoades, a long-haul trucker from Houston, was arrested in April 1990 on I-40 about 50 miles north of Phoenix under highly suspicious circumstances, as described in a 2012 article by Alex Hannaford in *The Texas Observer*: "A state trooper, who thought Rhoades

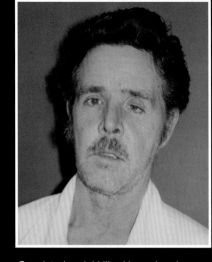

Convicted serial killer Henry Lee Lucas. (Courtesy of Texas State Library and Archives Commission).

was parked dangerously, discovered a woman inside the truck, alive, but shackled to the door. She had welts on her body, cuts on her mouth, and a horse bridle secured around her neck. In Rhoades' briefcase, investigators found alligator clips, leashes, handcuffs, whips and dildos. He had told his latest victim—the woman in the cab—that he had been torturing women for 15 years as he crisscrossed America by highway."

But I-45 between Houston and Galveston, named by *Cold Case USA* as "the bloodiest stretch of highway in America," is perhaps the most infamous site of serial highway murders, starting in the 1970s.

I-45 Murders

"From 1971 into the 1990s, more than 30 bodies—all of them schoolgirls or young women— have been found dumped along a 40-mile stretch of Interstate 45 between here [Houston] and Galveston Island, an area that's become known as the Killing Fields," wrote Alex Hannaford in *The Texas Observer*. The victims in the I-45 cases were mainly quite young, ranging in age from ten to twenty-three, and disappeared while they were out alone, either jogging, or en route to visit a relative, or on the way to a local store, or just walking in her own neighborhood. Some simply vanished, while others were discovered sexually abused and left dead in a remote spot. Frequently their remains were too decomposed to leave any clues or DNA for assistance in identifying their killer. There are many signifiers that point both to one chief murderer and to multiple serial killers who have hunted for their prey in the same I-45 areas—the stretch of road also known as 'The Highway of Hell,' due to its high rate of traffic accidents.

According to KPRC-TV in Houston, "Houston ranks No. 1 among US cities thought to have the most victims of human trafficking. The rank comes from new numbers released on the total calls made to the National Human Trafficking Resource Center tip line. The crime is so underreported that FBI agents say calls to a national tip line indicate the size and location of the problem." The trafficking comprises everything from forced prostitution to kidnap and slave labor, and is particular to the Houston area because of the city's function as a hub of ports and highways that give access to every other destination in the country—a system very handy to Mexican and Honduran traders bringing victims up from across the border and through the Gulf. Among states, Texas ranks second in human trafficking, just behind California.

Drug Smuggling

The drug cartels of Mexico and Central and South America have long enjoyed and employed the fine roads of Texas in their commerce. During the recent increase in oil industry activity, especially in the Eagle Ford Shale area where policing has grown more difficult due to the higher transient population, many vehicles disguised as service trucks, school buses, utility trucks (water and/or electrical), and even TxDOT trucks have ferried loads of illegal narcotics through the network of roads under cover of the heavier industry traffic moving along them. As reported by National Public Radio station KUT in Austin in 2014, "Because there are so many different companies, and so many different trucks going through that area, it provides a sort of way to blend in if you will," *National Journal* writer Ben Geman tells the *Texas Standard*'s David Brown. "Essentially what's happening is you've got smugglers who are stashing marijuana, or other drugs, in trucks that are either 'cloned' to look like one of the industry trucks, or some type of truck that seems to fit right in driving around on these ranch lands."

A) "Cloned" TxDOT truck used to ferry illegal narcotics and B) bales of marijuana discovered inside the truck.

The Transmountain Road travels through the Franklin Mountain State Park. Note the scenic overlook in the bottom right of the photo, offering motorists a chance to look back toward El Paso. (Michael Amador/TxDOT)

8

The 1960s, 1970s, and 1980s

The subject of Beautification is like a tangled skein of wool. All the threads are interwoven—recreation and pollution and mental health and the crime rate and rapid transit and highway beautification and the war on poverty and parks . . . everything leads to something else.—Lady Bird Johnson, diary entry, January 27, 1965

You get the water off and the rock on and you've got a highway.
—Engineer-Director Mark Goode, State Department of Highways and Public Transportation

If a national prize existed for the state with the most unruly outline, Texas would certainly be a frontrunner, if not the indisputable winner. Not only does the awkward splay resembling a lopsided four-pointed Lone Star with two tips broken off hold multiple terrains and ecologies within its borders,

D. C. Greer, 1960. (TxDOT Library)

but the rippling boundaries of two rivers and the Gulf are offset by the clean, sheared lines of the squared-off Panhandle—a contradiction to any neat geometry. And of course, the regions within that outline contrast as sharply with each other as if they lay in countries separated by entire oceans, as indeed, in previous geological eras, they did.

The Challenge of Opposites

On two of the most far-flung points across the land, in cities located 745 miles, several cultures, and at least three climates apart, the problems of easy passage and traffic congestion have historically looked very different. The mountainous desert topography of El Paso has attracted a much smaller permanent population than the lush tropics and steamy bayou flats of Houston, and the West Texas isolation of that finial jutting out between New Mexico and Mexico could scarcely be more unlike the port, ship channel, and bustling industrial hub of the second. But Fort Bliss, the army base situated at the edge of El Paso's outskirts, happens to be the US Army's second-largest installation, covering an area of 1,700 square miles and supporting 167,416 inhabitants. This number, added to the local El Paso population of 674,433, might conceivably present some transportation challenges, especially given the fact that a rugged and gullied mountain range bisects a large portion of the city from north to south.

It is sometimes easy to overlook El Paso's importance in the scheme of the pan-Texas system, precisely because of its inconvenient geographical placement and distance from any other large American population center. From early settlement on, the city's location has served as a prime military spot—first as an outpost, then as a post—and has required a workable road ingress and exit. And in the late nineteenth century, and continuing up through the 1920s and early 1930s, El Paso became a refuge for sick people: tuberculosis patients poured into its precincts from all over the country, seeking the healing benefits of its dry mountain air. Several hospitals and sanitoria were established there to treat people with the disease. Other victims who could not afford proper care or housing made their way there anyway and sought to enjoy the same healing conditions by camping out in a tent city on the slopes of Mount Franklin, until finally certain religious organizations and independent physicians built more hygienic quarters to serve them. At least one of the early clinical facilities has since developed into a major medical center.

The conduit El Paso provides does not end in Texas but pushes on through New Mexico and Arizona to California. The Texas Highway Department, though, has never forgotten or underrated El Paso's significance. After all, the El Paso District is the largest in the state, which is the reason that Dewitt Greer placed a strong chief architect in its district engineer position

in 1963. Joe Battle had begun working in El Paso in 1937 as a member of a Texas Highway Department surveying crew while he was still an engineering student at the University of Texas. After earning his degree in 1940, he immediately rejoined the department (his tenure commencing at the same time as Greer's administration) and remained there for another forty-five years.

Previous to his 1963 return to El Paso, he had acted as the assistant engineer for the Houston Urban Expressways Division, working with J. C. Dingwall to build the Gulf Freeway and others; with such dual experiences, he was able to understand the demands of both Houston and his new home. He served as El Paso's district engineer for twenty-eight years. During that period, his supervision guided the design and construction of virtually every major road in El Paso—which included (as well as other, smaller projects) the Border Highway, the final Texas segment of I-10 (completed in the 1990s), and El Paso's monumental Transmountain Road. Of these, the last has proven to be the focal jewel on El Paso's map.

Plans for the Woodrow Bean Transmountain Road, or Highway, had evolved through an intriguing mixture of hopes and engineering feats over twenty years prior to its construction, when the barrier presented by the Franklin Mountain Range to increasing truck and car traffic had continued to hamper any efficient access from one side to the other. In 1959, a survey

Under Joe Battle's guidance, the El Paso highway system evolved into a modern network evocative of the cultural and geographical surroundings. (Kevin Stillman/TxDOT)

The Transmountain Road (Loop 375) joins northeast and northwest El Paso, which were formerly separated by the Franklin Mountains. Opened 1969 to connect I-10 on the west with El Paso's World War II Road 11 on the east. (Jack Lewis/TxDOT)

El Paso's Loop 375 looking west toward the Franklin Mountains. (Benard Stafford/TxDOT)

was commissioned at a cost of $65,000 to establish the most acceptable passageway. But the results remained inconclusive, depending on the interests involved; internal conflicts arose, with many arguments over routes and the prices for executing them. One of these even included a fistfight, conducted in 1961 in Austin, when the mayor and the county commissioner "slugged it out" over the route of an alternative way around the mountains. Then, because the routes crossed army land, the military entered the planning fray. Eventually, part of the Castner Artillery Range at Fort Bliss was moved to make way for the asphalt strip.

A description of the finished product that had been directed by District Engineer Joe Battle, his assistant Max Moore, and several other Texas Highway Department authorities; excavated by H. B. Zachry of San Antonio (who removed 5 million cubic tons of rock and dirt and prepared the roadbed); and paved and sealed by the contracting firm of Hansen, Anderson and Dunn, can be found in a contemporary article from the *El Paso Times*:

> An old dream came true today when city, county, Texas Highway Department and Historical Society officials cut a ribbon opening the $5.5 million Trans-Mountain road across the ragged spine of the Franklin Mountain range.
>
> The spectacular road is deceptive in its easy seven per cent grades, its roomy 150-foot width, its barrier-less wide median, and its tour-lane driving surface. It actually rises from the east at War Road 11, where it is 3,900 feet above sea level, and tops off at Smuggler's Gap at 5,250 feet, the gentle grade and smooth six per cent curves belie the fact that a lot of motor labor is being done to move vehicles up to almost a mile in the sky.
>
> The crowd at ribbon cutting ceremonies was outlined by the rugged mountain range, which rises to a 5,440-foot height.
>
> Overlooking the gap, where construction crews had to tear away some 200 feet of rugged terrain, is the majestic Mt. Franklin peak. . . . Now, at last, El Pasoans and their visitors will have a different way to move from one side of the mountain to the other.

In October 1991, one week after his retirement from the State Department of Highways and Public Transportation (rechristened TxDOT the previous month), the El Paso City Council voted to name the newly constructed 11-mile Loop 375 linking Montana Avenue on the city's east side to Transmountain in the northeast for the district engineer who had accomplished so much: Joe Battle Boulevard.

Starting with the first segment of the Gulf Freeway that opened in 1948, the expressways and freeways of Houston began to crisscross the city and its suburbs in comparatively rapid succession. The first section of Loop

Houston's Gulf Freeway in 1954. The stadium in the photo was home to Houston's minor-league baseball team and a practice facility for the Houston Oilers. (TxDOT Photo)

610 opened in 1952; the first section of the Baytown–East Freeway opened in 1953, as did the first section of the Eastex; the first section of the Katy, soon to be rechristened I-10, opened in 1956. Through the next five decades a long string of other urban roadways followed: the Downtown Split, the Southwest Freeway, Northwest Freeway, La Porte Freeway, South Freeway, Spur 5, Tomball Parkway, Crosby Freeway, Grand Parkway, Baytown Freeway, Decker Drive—and the four toll roads that also now hasten and somewhat mitigate the ever-increasing traffic of a growing metropolis. In fact, since World War II, Houston's population has swelled at an astonishing rate, placing a strain on the ecosystem and the transportation options of its 655-square-mile area; currently the fourth most populous city in the United States and the fastest in growth (the second is Austin), its residential numbers increase by an estimated six hundred per day. (According to U-Haul, Houston has been the nation's number-one moving destination for those renting U-Haul's equipment.) Today one 23-mile portion of its I-10 Katy Freeway is literally *one of the widest roads in the world*; at Beltway 8, it stretches twenty-six lanes across.

There are twelve main lanes—six in each direction—eight feeder lanes, and six managed lanes, which carry mass transit and high-occupancy vehicles during peak hours and are made available to single-occupancy vehicles for a toll fee during off-peak periods.

For a municipal district that could contain the cities of New York, Washington, Boston, San Francisco, Seattle, Minneapolis, and Miami put together, where ethnic diversity is reflected by the fact that at least ninety languages are spoken daily in its streets, the question of infrastructure is crucial in the way it keeps this jigsaw society on the move and weaves it together.

When reconstruction was complete, the new Katy Freeway was dubbed one of the widest highways in the world. (Smiley N. Pool/Houston Chronicle. *Used with permission*)

Several of the most prominent features on the Houston transportation scene are, of course, bridges, the inevitable result of a salty, soupy environment. One in particular stands out these days. But before its existence—before the beauty of its cable stays laced the sky above the Houston Ship Cannel—there was a tunnel. The Baytown–La Porte Tunnel—the first tunnel in the Texas highway system—opened in 1953, its heavy steel and concrete tube encasing a space beneath the Houston Ship Channel in a northeast-southwest line for a length of 4,110 feet. It replaced the Morgan's Point

US 75, now I-45, in Houston. (TxDOT Photo)

Loop 610 at the Southwest Freeway on the west side of downtown Houston, 1964. (Jack Lewis/TxDOT)

Aerial view of Morgan's Point Ferry on Texas 146, April 1939. Ferry service was established between La Porte and Baytown (Goose Creek) in the late 1800s. The service continued until the Baytown Tunnel was opened in the 1950s. (TxDOT Photo)

Interior construction of the Baytown Tunnel. (TxDOT Photo)

Worker places tiles on the interior of the tunnel. (TxDOT Photo)

The Baytown Tunnel open for traffic. (TxDOT Photo)

The Baytown Tunnel contained 164,000 square feet of white ceramic tile and 16,500 square feet of terra-cotta. It was washed weekly in a bath of thirty thousand to sixty thousand gallons of water from this high-pressure tunnel washer. (TxDOT Photo)

Ferry, linked the two Houston suburbs for which it was named, and connected State Highways 146 and 225 on the southern end to State Highway 146 and Spur 201 on the northern one. A masterful example of engineering, it stood up well to the water pressure and natural deterioration such conditions cause for most of its forty-two years of constant daily use, but already in the 1970s it was far exceeding its capacity of twenty-five thousand vehicles per day. Its prefabricated 300-foot sections of 1-inch-thick steel cylinders were walled in concrete 2 feet thick, for a total diameter of 35 feet; the flat floor roadbed of the tunnel covered a ventilation shaft.

By the early 1990s the heavy traffic had taken its toll on these materials, and their worsening maintenance problems prompted the public to demand the tunnel's replacement with the bridge that had been in various stages of planning since 1967. The State Department of Highways and Public Transportation contracted Williams Brothers to build the magnificent cable-stayed Fred Hartman Bridge, which was completed and opened at the end of September 1995, and the tunnel's subsequent slicing up and extraction two years later to make way for a deepening of the Houston Ship Channel provided an interesting test of ingenuity for TxDOT and Williams Brothers engineers, an engrossing spectacle for onlookers, and another first for Texas: never before had an underwater vehicular tunnel undergone removal in the United States.

Construction of the Hartman Bridge in 1994. (Gay Shackelford/TxDOT)

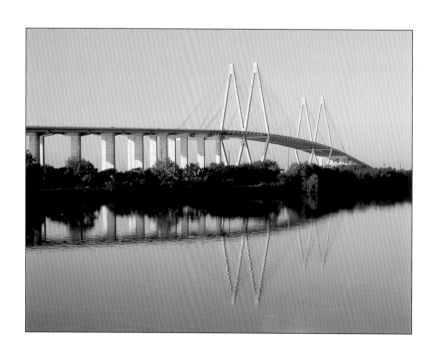

The Fred Hartman Bridge or Baytown Bridge is a cable-stayed bridge spanning the Houston Ship Channel. The bridge carries 2.6 miles of State Highway 146, between Baytown and La Porte. The bridge, named for Fred Hartman (1908–91), editor and publisher of the Baytown Sun *from 1950 to 1974, is the longest cable-stayed bridge in Texas. The construction by Williams Brothers cost $117.5 million. (Kevin Stillman/TxDOT)*

The Baytown Tunnel was removed in
sections by Williams Brothers to allow
deepening of the Houston Ship Channel.
(Kevin Stillman/TxDOT)

Fred Hartman Bridge. (Stan Williams/TxDOT)

The Discovery of Inner Space Caverns

James W. Sansom Jr., Texas Highway Department geologist

Famous last words: "Take care for you might encounter some caves while drilling this location."

In the spring of 1963, I was employed as the geologist for the Bridge Division of the Texas Highway Department. The Austin District requested a drill rig to make core drillings for the bridges and overpasses of the I-35 bypass south of Georgetown. The core driller discovered a large cavern at the first drill location. The drill bit had dropped 25 feet from where he drilled into the roof of a cavern to the floor of a large room which is now called the Discovery Room by the Inner Space Cavern owners.

Department geologist Jim Sansom prepares to ride a drill shaft down into the cave discovered during construction of I-35 near Georgetown in 1963. With Sansom is Jim Cole, core driller for the department's Austin District. (Jack Lewis/ TxDOT)

A 24-inch-diameter hole was drilled into the cavern so we could map the extent and condition of the cavern relative to foundations for the planned overpass. Jack Bigham, Bill Schultz, Lawrence Schultz, Horace Hoy, and I were lowered down through hole by a make-shift stirrup on the end of the auger rig. Jack Bigham was the first to enter the cavern and I was the second. We explored the more accessible portions of the cavern and were surprised of its size. District personnel surveyed the cavern and mapped the major portion of it that was underneath the proposed overpass. Highway department bridge engineers determined that the 33.5 feet of competent limestone between the surface and the cavern was adequate to support the planned overpass. Dr. William W. Laubach, the landowner, received permission from the highway department to develop the cavern beneath the interstate. Inner Space Cavern opened to the public in June 1966 and continues operation today.

Geologists and engineers descending through 33 feet of solid lime-stone. (Jack Lewis/TxDOT)

Highway Department engineering aide Bill Johnson beside an elaborate "curtain" stalactite that was among a profusion of subterranean formations. (Jack Lewis/TxDOT)

Beautification

One of the most famous federal accomplishments related to highways in the 1960s arrived during Dewitt Greer's final years as state engineer and had a decidedly Texan flavor. In 1965, President Lyndon B. Johnson pushed forward his intent to see a bill passed by Congress that would make provisions for highway landscaping across the country (as Texas had been practicing since the Jac Gubbels years of the 1930s), eradicating eyesores such as outdoor advertising and the giant billboards marring both sides of the highways, eliminating and/or controlling all junkyards abutting interstate roadbeds, and conserving and restoring the natural environments so often destroyed during interstate construction. The Highway Beautification Act of 1965 had a number of fierce opponents, almost all of them Republicans. Some of the objections had to do with dedicated federal highway funds being drained to promote these goals—especially the monies that would be diverted from maintenance and construction into the landscaping and conservation projects. By far the greatest controversy, however, entailed the restriction of signage. It seemed the dissenters felt that outdoor advertising, no matter how big and obnoxious, should be an American right, conferred as an adjunct of the free-enterprise system.

Wildflowers along TX 71 between Brady and Llano. (Randall Maxwell/ TxDOT Library)

(Michael Amador/TxDOT)

As the chief advocate of the bill, Lady Bird Johnson received much criticism and mockery from the opposition. She remained unfazed. Like the two longest-serving and most foundational Texas state engineers, Claudia Alta Taylor "Lady Bird" Johnson had been reared in rural East Texas. She grew up amid the pine forests, the fields of wildflowers, the waters and cypresses of Caddo Lake—surroundings that filled her days with a consciousness of the natural world and sensitized her forever to the human need to connect to its beauty. She saw the job of cleaning up our roadways as a social tool that would affect everyone in a positive way, improving the hearts, minds, attitudes, moods, and self-respect of citizens across the nation. But achieving what she and her husband both considered at least a partial remedy for the ugliness humans had created along US roads was a daunting task; support had to be wooed and won even from their own Democratic Party members.

On October 22, 1965, the same evening the bill came up before Congress for its final vote, a large formal event was being held at the White House, and many congressmen and their spouses were invited. The wives began to arrive, gathering on the porch before the White House doors in their evening dresses, smoking cigarettes, growing more perplexed and impatient as they stood hovering in the evening chill, waiting either for their husbands

(LBJ Library)

to appear from the Capitol, properly attired in tuxedos, or for the doors to hospitably open. But President Johnson did not permit his guests to enter. Instead, he kept them dawdling outside while he telephoned US Representative Jim Wright, who was attending the House of Representatives session before joining the festivities. According to Wright, Johnson instructed, "Do not bring those members here until you've passed the Highway Beautification Act." No one would be allowed into the hall until he got his way. When the bill passed a short time later, its voting division lay along strict party lines.

At the signing ceremony, Johnson announced, "This Bill will enrich our spirits and restore a small measure of our national greatness. Beauty belongs to the people. And so long as I am President, what has been divinely given to nature will not be taken recklessly away by man." Then he handed the first signing pen to his wife, Lady Bird, and kissed her on the cheek. Thus, the concepts Johnson had pursued as a congressional aid running the Texas branch of the National Youth Administration, assigning and overseeing his workforce of boys and young men to build pleasant, verdant rest areas beside the highways of his home state, were extended into a sweeping program intended to make that pristine verdancy once again available to the whole country. The precedent set during the Great Depression by Gibb

Interior Secretary Stewart Udall, Lady Bird Johnson, and others tour Big Bend National Park. (LBJ Library photo by Robert Knudsen, April 2, 1966)

The entrance to the LBJ Ranch in Stonewall bears the designation Ranch Road 1. (Jack Lewis/TxDOT)

After the Johnsons returned from Washington, D.C., to Texas, the former first lady worked with the Highway Department to establish the highway beautification awards to recognize employees for their work in beautifying Texas highways. The awards ceremony was held at the LBJ Ranch and carried cash awards for the winners. The first ceremony was held on October 2, 1970, and the former first lady and president were in attendance to celebrate the winner, Joe H. Derrick, here pictured with his wife and Lady Bird. President Johnson quipped after the presentation: "I come over here today because I've lived with Lady Bird since 1934 and I've never seen her give away $1,500. I want to ask Joe Derrick how he does it." (Herman Kelly/TxDOT)

Gilchrist, Jac Gubbels, and the two Texas Highway Department employees who had first demonstrated the initiative to donate and develop their own little wayside parks, helped make a lasting mark on the national environmental awareness. As a result, our highway journeys grew more beautifully scenic year by year.

The Automation Revolution

For decades, the surveyors and engineers of the Highway Department plotted roads across topography using the same methods. First they trudged over hills and gullies, camping out, riding on long horseback journeys through almost-impassable terrain while they took their painstaking 100-foot measurements by hand. Then they applied tedious, long, inch-by-inch fans of angled calculations to come up with the best correct lie of the prospective roadbed. But in the 1960s, for the first time, all these steps changed.

During the early 1960s the revolution commenced that altered the engineering procedures (and other functions) of the department. Up until that time (and moving into the 1970s), the hand-drawn and hand-calculated designs produced by engineers had been the staples; this was also true in the financial management areas of the agency. But then in 1960 and 1961 computers entered the picture. Eventually, these would pave the way for the first entirely automated designs and transform the structure of the department in the field and across the state, enabling engineers to plot roads across difficult topography; instead of foot-slogging surveys, aerial photos could be taken that then fed information onto large paper sheets—nicknamed "saddle blankets" by their operatives—and shipped to Austin. Staff members then entered the data by filling out punch cards and feeding them into the enormous room-filling IMB 1401, a seven-unit data-processing

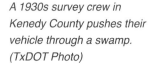

A 1930s survey crew in Kenedy County pushes their vehicle through a swamp. (TxDOT Photo)

center acquired in 1961 and installed in the downtown Austin Highway Department building's basement. This computer chiefly served the Motor Vehicles Division; automation of vehicle registration and titles simplified what had formerly been a laborious bureaucratic coordination. But soon a clever programmer, Larry Walker, developed code to utilize the computer during its free time and put it to work performing tasks beyond the first-use spectrum, analyzing all kinds of other data for reports, from the best road-surfacing materials for a specific area and climate to how many miles of unpaved county roads serviced enough traffic to warrant paving them. The data computations usually took anywhere from fourteen to twenty-four hours for the machine to accomplish, especially after it was also employed to develop designs. A day, sometimes two, would pass, as the computer ran through all of an engineer's analysis for a road design and produced a large roll of paper displaying all aspects of the putative results.

Sometimes, however, a punch card might be deficient due to human error; someone had missed a single computation, line of code, or column when entering the data. When that happened, the computer could spend the day producing blank, inkless rolls of paper with perhaps one tiny square appended to the end or jagged lines spraying out in different, nonsensical directions—all because of that single omission. Then the engineer waiting eagerly in his district office somewhere across the state for the finished paper roll to arrive with his designs on it would be crestfallen to discover the vast blank tundra of his results, and the painstaking, days-long process had

Advances in automation during the 1960s aided in many areas of highway engineering and development. Tommie Howell, assistant head of the photogrammetry section, checks a new application of photogrammetry. Draftsmen Pat Hoffman and Tom Johnson use topographic maps plotted from aerial photographs to construct a scale model of a complex area of interchanges and bridges in Houston. The model will be used to clarify points in a court trial. (Robert McCarty/TxDOT)

Ben Alley and Max Ulrich, Austin District employees, assist model builder Pat Hoffman (center) build precision scale models used to illustrate prospective projects to the public, to highway contractors, and to juries hearing right-of-way location and condemnation cases. (Herman Kelly/TxDOT)

Modern applications have grown exponentially; here an inspector uses a tablet on a US 281 construction site in Marble Falls. (Will van Overbeek/ TxDOT)

to start over again: the search for the missing piece of data, the shipment to Austin, the card punching, the computer translation, the week's waiting, the arrival, and the bitterest disappointment an engineer can face: the prospect of a lot of wasted time and effort. Later, in 1967, an IBM 360 computer with a magnetic iron-core memory, the most common mainframe system in use at that time, replaced the 1401. In 1970, IBM 370s with integrated circuitry instead of iron cores replaced the IBM 360s (which could heat up to such a degree that they sometimes caught fire). It performed far more quickly, incorporated more data into its memory banks, and was also used in coordination with the Department of Public Safety.

By the time Marquis Goode and, after him, Raymond Stotzer inherited the mantle of directorship, further advances in technology had eliminated punch cards and enabled near-instant computations that would eventually be performed on small personal devices. Bookkeeping and balance sheets entered a new, more efficient era. New aids also appeared to refine other departmental processes, such as satellite photography and the tight location of geographical coordinates through GPS, and the Internet, expediting communication and data transmissions. The old slide-rule days were over.

Leah Moncure, the first woman engineer hired by the Highway Department in 1938. (TxDOT Photo)

In the late 1980s, what was then known as the Automation Division paid approximately $18,000 per annum to employees who could then be lured away to companies like IBM for double that salary. It became a serious problem to retain skilled high-tech personnel in the face of such attractions. Hubert Henry, who headed the division at that time, met with Executive Director Marquis Goode to express his concerns. Marcus Yancey described the subsequent conversation.

"Hubert, I just don't understand why we're losing them," said Goode when he heard just how high was the drop-out rate. The Texas Highway Department was unaccustomed to seeing staff members depart for other pastures—in fact, throughout its history, such behavior had remained almost unheard of. Lifetime allegiance and devotion, not only to the state but to the highway family, had for decades been the norm, because the department both earned and rewarded those attributes.

"Boss, it's just a matter of money, for most of them," Henry replied. "But we've got a lot of challenges and we choose a lot of people because of that. And I have some suggestions."

"Well, just what kind of people are you looking for?' Goode asked.

"Weird," said Henry. In other words, Yancey explained later, they had to think out of the box.

"Well," Goode said, "I don't know that we want any of those kind of people."

"You've got a lot of them right now out there, and that's what's making us go. If you're going to get people [to stay and perform well], that's what they have to be."

At the console of the 1604A Control Data Corporation computer complex in 1966, Ted Ball (right), head of computer operations, and Automation Division Engineer/Director Hubert A. Henry confer with Dean Taft, computer operations supervisor (seated). At the right, Jim Daude adds tape to one of the four magnetic tape drives. (Robert McCarty/TxDOT)

They needed to have a personal loyalty and commitment to being "weird" that they valued above the financial rewards as well. In addition, their weirdness needed to receive appreciation, respect, and gratitude for its own unique contributions.

Changing of the Guard

Just as profound as the technical revolution that swept through the department in the 1960s was the change in leadership. After nearly twenty-eight years at the helm of the department, Dewitt Carlock Greer stepped down into retirement on January 31, 1968. Soon afterward, he accepted a position as professor of engineering practices in the Civil Engineering Department of the University of Texas at Austin. In 1969, Governor Preston Smith appointed him to chair the Texas Highway Commission—a move that *Texas Monthly* magazine later commented was "like picking Winston Churchill to be king of England." He served another twelve years on the commission, the first three of them as chairman, and while still on the commission saw the department's name change to the State Department of Highways and

Dewitt Greer accepts a copy of Lieutenant Governor Preston Smith's proclamation of Dewitt C. Greer Day in Texas, August 4, 1967. (TxDOT Photo)

Public Transportation (SDHPT) in 1975, when the agency merged with the Texas Mass Transportation Commission, and the title of his former position of state engineer change to engineer-director that same year. Although he did not meddle in the day-to-day functions of the department from his place on the commission, he continued to influence and, some claim, to actually run its larger policies and judgments—including exerting a strong preference for no rise in the gas tax, which remained a steady 5 cents per gallon from 1955 to 1984. (After his departure from the commission, the legislature raised it to 10 cents per gallon and then to 15 cents in 1986, both times at the persuasion of Highway Commission Chair Bob Lanier.) Greer died on November 17, 1986, at the age of eighty-four, leaving a legacy of honorable public service, trained successors, superb highway construction practices, and a formidable edifice of reliability, professional family bonds, and dogged loyalty that would live long after him.

In 1967, the department established a new program of recognition that acknowledged outstanding service, innovational thinking, and the greater levels of achievement by an employee. The Gilchrist and Greer Awards are delivered at Short Course, the annual conference and gathering of departmental management, administrative, engineering, and higher-echelon personnel cosponsored by the partnership between the department and the Texas Transportation Institute on the Texas A&M University campus at College Station. Starting in 1981, the DeBerry Award was added to the list, followed in 1990 by the Stotzer Award, named after Engineer-Director Raymond Stotzer.

Dewitt Greer had handpicked the man who followed him into the executive seat of the Texas Highway Department. His personally schooled assistant James Colin Dingwall, who had worked for the department since 1928 and served directly under Greer for ten years, possessed a trove of experience in urban expressway, interstate highway, and toll road design and construction by the time he stepped into the role of state engineer. As a young man, while working on his first job building levees on the Trinity River in Dallas, he had attended Southern Methodist University but had not continued there long enough to complete the full requirements for a degree. For that reason the wording of the Texas Highway Department Act specifying the necessary credentials for the chief position was revised from "engineering school graduate" to "registered professional engineer," an alteration that would lead to further unforeseen repercussions more than four decades later.

Meanwhile, despite Dingwall's lack of a degree, his qualifications, earned through forty years of excellent performance for the department, were not only well respected but regarded as almost impeccable. He also exemplified Greer's principles, which were also followed up in his two successors: Greer felt it essential for the department chief to have risen from

Three High Plains highway builders during an inspection tour of a Highway Department ferry. Left to right: Percy Bailey, then assistant district engineer in Amarillo; G. K. Reading, who retired as supervising resident engineer at Pampa; and J. C. Dingwall, then a resident engineer at Pampa. Dingwall followed Dewitt Greer as engineer director in 1968. (TxDOT Photo)

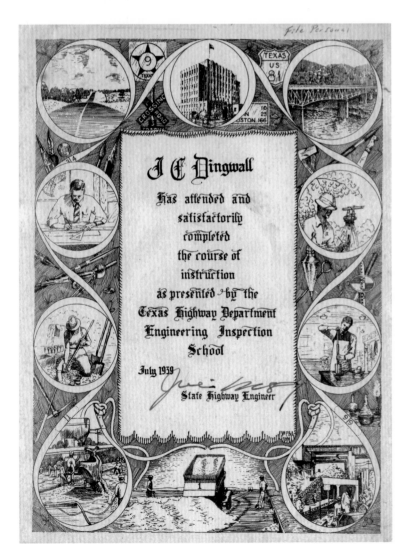

Dingwall's certificate to qualify him to serve as state engineer. (TxDOT Photo)

the very bottom and climbed upward through merit, experience, and hands-on fieldwork, no matter how menial, so that he or she would know all problems and what it took to find their possible solutions inside out—each phase and step in a project or a maintenance task, from the materials involved to correctly applying hot mix, from soliciting a project's funding to right-of-way purchases, from talking with county judges and other authorities to design to supervision of contractors and construction crews—and therefore understand what was required of all employees in every job tackled. Just as crucial was the trial-by-fire facet of this education, in which, as Marcus Yancey put it, the candidates would "get the rough edges knocked off them all the time—they had to learn how to take their licks, take that criticism, learn how to handle it and deal with it." Dingwall and the two men who followed him learned those lessons well.

Born in Comanche, Texas, to Scottish immigrant parents in 1907, Dingwall had joined the department as a lab assistant in Abilene right after leaving college and his levee-building stint on the Trinity. From there he moved to the High Plains—Pampa, Shamrock, Panhandle, and Borger—before joining the US Army during World War II, where he attained the rank of major in the Army Corps of Engineers and oversaw the construction of air bases in Laredo, Guatemala, Brazil, and the Azores. Upon rejoining the department at the war's end, he directed the construction of the Gulf Freeway in Houston and then became head of the Road Design Division at department headquarters in Austin in 1950, only to be "borrowed" from the department to help create the Dallas–Fort Worth Turnpike. After its completion in 1958 he returned to Austin to become Dewitt Greer's right-hand man. He, along with subsequent chief executives Luther DeBerry and Marquis Goode, was considered by all in the Texas transportation industry to be a member of "Greer's Triumvirate."

The Corruption Myth

According to Tom Mangrem, retired area (resident) engineer of Brewster, Presidio, and Davis Counties, "The rule to live by is you'll never do anything for personal gain. Everybody is going to mess up, but you'll be forgiven if you didn't do it for personal gain." Whether the Texas Highway Department's early susceptibility to the toils of other people's greed imprinted an expectation on the mind of the public or whether the sight of such a large agency naturally predisposes people to cynicism, the fact remains that many Texans through the years have automatically (with no evidence) assumed that a certain amount of looting or graft must take place within its coffers. No one could ever have faulted Dewitt Greer's zeal and honesty during his tenure as its chief. Greer himself was spotless; his frugal lifestyle and modest estate testified to his integrity. After his retirement, his professorship in no way guaranteed wealth beyond a steady academic income; nor did his appointment to the chair of the Highway Commission during 1969 through 1972 and his continuing presence on the commission until 1981 pay anything more than the most nominal compensation for his time.

Throughout those nearly twenty-eight years he had run a tight ship with such steely oversight and control that it would have been difficult, if not unthinkable, for any of his underlings to attempt a larcenous action, deliberate or not. The scalding lessons of the Ferguson era were never forgotten: no one would dare impugn the department's reputation by committing a wrongdoing that could bring it down to such a base moral level again. Greer had once fired a good friend and district engineer because the engineer's wife had invested in a firm that happened to be doing business within the department. All employees were even expected to make certain all their household and personal accounts got scrupulously serviced: "Don't ever let a creditor call here about your not paying your bills on time," a new young hire was warned.

Greer's successor, J. C. Dingwall, did not possess the same luxury of a decades-long dynastic authority when he assumed the state engineer's position in February 1968. One longtime Highway Department employee sympathized by remarking, "God is a hard act to follow." Certainly Dingwall, who would also serve as president of the American Association of State Highway Officials in 1972 (following in his boss's footsteps even in this, as Greer had been president in 1949–50), made sure to sustain and uphold Greer's policies and administrative techniques. But he had an array of issues to face that had not confronted Greer. Double-digit inflation affected funding, increasing environmental conflicts affected plans and procedures, minority and women's rights affected hiring and salary scales, and protracted delays affected the completion of the interstate highway system.

Under Dingwall's management, the department now maintained as many as nearly twenty-one thousand employees, more than ever before, as the push to complete the interstates and meet the federal deadline continued. And Texas was growing increasingly urban, thanks to the ongoing shifts of population aided by the new interstates and their easy flow into the cities. What had begun during wartime as the move from farms and ranches to industrial jobs generated by the war effort now had become a way of life that almost canceled out the previous generation's rural lifestyle. Progress meant the city and people wanted to catch up to progress. Although the Highway Department had been hugely instrumental in making this transformation not only possible but probable, its own workforce had to cope with the added pressures, the altering values, and the new desires and expectations for urban affluence, as much as did the rest of the citizenry. In such complex conditions, the temptation toward malfeasance—though a drastic departure from the trustworthy atmosphere and loyal family attitude fostered by Greer—could possibly chew away at the department's clean conduct record.

So Dingwall decided to install an internal auditing unit within the department that would detect any misbehavior, should it appear. It was the first system of its kind, according to Marcus Yancey, in any state and/or within any state agency. "Its sole purpose," said Yancey, "was to help management ferret out and find unaccountable uses of money, misuses of money, or just plain fraud. Or in some instances, [other] crime." Comprehensive investigations were carried out whenever the slightest hint of an abuse or violation came to its attention; employees were encouraged to report bad conduct toward themselves and others, such as sexual harassment or taking advantage of an employee's circumstances in any way. For some years, the unit operated quietly, discovering the truth about cases like the maintenance foreman in Sonora who encouraged employees (and possibly others) to line up their personal cars outside the maintenance facility gate at close of workday on Fridays and fill their gas tanks free of charge from the department's storage (on the morning after the auditor confirmed the report, a new foreman replaced the guilty one), or a certain East Texas district facility with a maintenance engineer who attempted, and sometimes succeeded, to intimidate female employees into sexual liaisons.

Proof of the unit's effectiveness gained broader public knowledge nine years after its establishment. Marcus Yancey, who at that time acted as assis-

Marcus Yancey, 1979.
(John Suhrstedt/TxDOT)

tant engineer-director and public spokesman for the department and during his career represented the department's interests during sixteen legislative sessions under eight governors, described the moment this occurred.

In 1977 we were going before the Legislature because inflation had increased to such a dramatic effect that the gasoline tax would no longer handle it. And we had to find another way. And the night before the session was to start the next day on the hearing on House Bill 3, as it was called, the Lubbock newspapers turned loose and said that an employee of the department, an attorney, A. C. Black, was now indicted for stealing at least a known amount of $245,000 from many minority people from whom we had bought right of way. What the papers *didn't* say was that it was our auditors that found that in the District. And when that hit the paper, the next morning it was my responsibility to go talk to both the Senate and the House on House Bill 3. And I said, "Before I begin, let me tell you the following history. And here are the reports." We went to them and said, "Here it is." And House Bill 3 . . . was passed because the Legislature trusted the Department. We wouldn't sweep it under the rug.

Federal funding during Dingwall's stint as chief executive became a hit-or-miss proposition, despite the full coffers of the federal Highway Trust Fund. Some of the monies' distribution was constricted due to the Vietnam War effort; the government was manipulating their release or withholding them altogether from the states that needed them to complete federally mandated projects such as the interstate system to try to better control the wildly inflationary wartime economy. The other biggest challenge Dingwall had to face at the same time was the torrent of red tape engendered by recent federal guidelines and initiatives—a spate of new regulations—that hampered any chance of meeting the interstate 1972 completion deadline. Issues previously ignored or unheard of swelled during the 1960s and escalated further with the launch of the 1970s. Environmental concerns, civic scrutiny, civil and women's rights and resultant personnel revamping, and more direct public involvement and feedback all now took their toll in the form of time-consuming bureaucratic demands and documentation. The deadline was to occur on Dingwall's watch, after all, and as it approached, he began to despair.

As a result of the additional reams of paperwork, the average lead time in which to bring a construction project from conception to bid letting, not even counting fruition, had grown from thirty-four months in 1956 to six years and five months by September 1971. In his frustration, Dingwall, who was also at that time president of AASHO, constructed a flow chart with the

Luther DeBerry and J. C. Dingwall examine the project development flow chart. (TxDOT Photo)

help of Marcus Yancey and several members of the Highway Design Division, listing and demonstrating all the things that could go right or wrong within the process during that almost-seven-year time frame, including the filtering parade of paperwork inching through the appropriate federal offices. By the time it was finished, the chart was 15 feet long.

Dingwall had it photographed; then he shared it with his colleagues at AASHO, who in turn submitted it for newspaper publication in states throughout the country. When Dingwall was summoned to Washington alongside two other administrators to testify before a congressional committee on the red tape problem and its delaying powers, a New Hampshire congressman told him that Texas was the cause and root of the hearings; he himself had taken a copy of the flow chart to the Subcommittee on Investigations and Oversights as evidence of the snafus and their effects. The subcommittee, in turn, chided the Federal Highway Administration (FHWA) after the hearings for imposing tighter restrictions than the statutes required.

At the end of January 1973, J. C. Dingwall retired from the department at the age of sixty-five and moved with his wife back to Comanche—a sweet full-circle homecoming after at least half a century spent far afield. He had accomplished a sound period of management, although, as he retorted to environmentalists who predicted that by the 1990s, Texas would be "covered with asphalt," there was virtually no new construction being done, except for paving through suburban subdivisions. "All the work we do now is to upgrade roads," he said. "The road is already there. We're not adding any new miles." He had often quoted one of Dewitt Greer's favorite mottos, "Don't ever let the sun set on a pothole," and then borne it out in asphalt and cement. He had also approved a memo within the department that would made a big difference to quite a number of its employees: he okayed the decree officially negating the unwritten dress code that denied female personnel the right to wear pants to work. The next person to fill the state engineer's spot, Bannister Luther DeBerry, navigated the department through some of the roughest periods and trials it had encountered since the early years of the Ferguson feeding fest.

Pants suit memo. (TxDOT Photo)

Tourism

In 1974, the publication that thrives under the management of the Travel Division issued its first copy in its new form as the official travel periodical of the state, as designated that same year by the legislature: *Texas Highways* magazine. Although the actual magazine had been published since the first volume was printed in 1953, this version with its beautiful color photography and features on unique and interesting Texas destinations, novelties, and even bird and wildlife portraits would become even more popular with the general public and add to the department's luster as a tourism-promoting mechanism.

The old travel information centers founded in the 1930s by Gibb Gilchrist's administration had evolved into sophisticated hospitality stops replete with literature that included a *Texas State Travel Guide*; the official state map as well as maps of cities; county grid maps; and many brochures, pamphlets, booklets, and offerings of local and statewide sights, points of interest, lodgings, dining spots, and recreational facilities of all kinds. The new magazine published, and still publishes, articles on historical trails, fiestas and celebrations, seasonal natural attractions in different parts of the state, museums, ethnic cultural settlements with their backgrounds and

In May 1974, Texas Highways morphed from an internal employee magazine to what became a multi-award-winning publication celebrating travel and tourism in Texas. (TxDOT Photo)

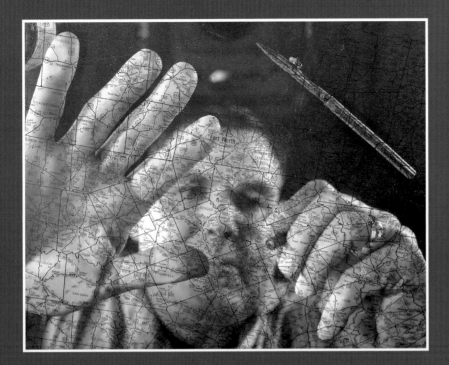

Touching up the 1967 official state highway map before it goes to press is J. D. Eilers, Travel Information Division draftsman. (Hugh Pillsbury/ TxDOT)

Frank Lively, editor of Texas Highways *magazine, 1962. (Hal Stegman/TxDOT)*

traditions, music festivals, theatrical productions, regional food dishes, crafts, and even alcohol tastings—a range of subjects as wide as Texas itself. The Travel Information Division has since also evolved to house several other but related functions: cleanup programs, reports on highway conditions, and the TxDOT Photo Library (an invaluable resource available to the public and for business use—a compendium of historical and current images illustrating the entire record of the Texas Highway Department's achievements and culture—now transferred to the Communications Division).

The Second of the Greer Triumvirate

Luther DeBerry, yet another Northeast Texas native to reach the top post in the Highway Department, was born in Bogata in Red River County, less than 100 miles from Gibb Gilchrist's birthplace of Wills Point, less than 50 miles from Dewitt Greer's hometown of Pittsburg, and about 100 miles northwest of Karnack, where Lady Bird Johnson grew up. In other states, this proximity might seem more far flung, but in Texas, it suggests a common cultural background quite different from those of the state's regions lying hundreds of miles to the west, south, or angled up through the Panhandle—almost, in fact, a neighborliness. Since one of the factors these county areas shared was cotton farming as the main agrarian pursuit, all three of the East Texas executives understood the need for good rural roads during bad weather that turned the sticky red clay soil of the district into a stew, and none more intimately than Luther DeBerry.

Luther DeBerry. (John Suhrstedt/ TxDOT)

The son of a cotton farmer, DeBerry began a roads career at an early age, building tracks through that soil across the family farm, with mule power driving the equipment. From there he determined to become an engineer and took a job as a rod man on a Texas Highway Department survey crew in 1934 while still an engineering student at the University of Texas. This was during a time when all employees not laboring on a road crew were expected to wear khaki pants, long-sleeved white shirts, and bowties, no matter how high the temperature climbed on a summer day. But the sartorial prerequisite failed to daunt DeBerry. After graduation, he joined the department on a more official basis, working his way up through the positions of resident engineer at Emory, senior resident engineer at Greenville, assistant district engineer at San Antonio, and district engineer at Lufkin and Dallas.

By the time DeBerry graduated from his position as Dingwall's acting assistant to the top executive post, he had learned enough wisdom and practical skills from his mentor to divide the department into two functional parts and to name two supervisors to oversee them: Marquis Goode for the operational side and Marcus Yancey for the administrative side. He also had a buffet of new predicaments to confront beyond the completion delays of the interstate system. Only a few months into his term of service, the oil crisis of 1973 went into full effect, triggered and then exacerbated by OPEC's (Organization of the Petroleum Exporting Countries) embargo on oil exports in response to the US role in supplying arms to Israel for its defense during the Yom Kippur War. Also included in the embargo were the Netherlands, the United Kingdom, Canada, and Japan. American dependency on Middle Eastern oil at that moment turned out to be a vulnerability with consequences few could have foreseen. Hardest hit was, of course, the transportation industry. And because of Texas' traditional status as an

(TxDOT Photo)

oil-proud state, the annoyance within its borders fused into defiance, indignation, and angry bigotry (all of which reheated a few years later during the second energy crisis of 1979, when the shah of Iran fled the Iranian Revolution, Iran's oil production fell to severe limitations, and the Ayatollah Khomeini suspended its exports).

During the 1973 crisis, with the skyrocketing price of gasoline followed by shortages, the federal government determined that strict measures must be taken. Service stations closed on Sundays. Congress passed a 55-mile-per-hour speed limit for the entire nation, backing it with the warning that states unwilling to comply would lose all their federal highway aid. Motor fuel tax collections dwindled; cars were left deserted in driveways unless a trip became imperative. The interstate program stalled on the national level. And within the State Department of Highways and Public Transportation, maintenance faltered to a near halt, and a hiring freeze ensued. Inevitably, Texans resented the 55-mile-per-hour national speed limit imposed on them. But for the state legislature to match the federal mandate by effecting a similar statewide limit lay beyond most of the public's tolerance, and lawmakers doing so feared for their political lives. After two special session attempts, they delivered the responsibility to the Highway Commission instead.

For most of his seven-year tenure, DeBerry had to campaign for more office space in which to house the department in Austin. At least two building projects were either canceled or vetoed by the state government. Despite the hiring freeze, the attrition rate, and the whittling down of total staff numbers throughout the 1970s, work space was at such a premium in the department's official buildings that it had to be rented in locations scattered all over town. Then, starting in 1975, as the cost of work and equipment climbed ever higher and the gas tax, fixed as it was, failed to provide enough fiscal support to pay the prices, the department had to resort to cutbacks. Bid lettings ground to a halt; right-of-way acquisitions were banned; worst of all, the need for actual layoffs became paramount. When DeBerry ascended to what was soon to be retitled the engineer-director seat, the department was sustaining around 19,500 employees—1,000 fewer than the previous high number under Dingwall but still a substantial group. DeBerry bore the painful burden of issuing the order to division heads and district engineers to reduce those positions by more than 5,000, skinning the workforce down to 14,000. "That was the hardest thing I had ever done," he said later, "to sign that letter." And it was the first time in the department's history that such a step had been necessary.

Another difficult chore, and one DeBerry had to implement within the department among lifetime career hands for whom the move was extremely unpopular, was the restructuring of the agency. The new title—the State Department of Highways and Public Transportation—and all it implied did not sit well with many of these longtime administrators and employees, who felt it posed the threat of potential fund dilution. Highway monies might get diverted to other systems, such as rapid-transit rail. DeBerry and the commission pointed out that they were in the transportation business rather than merely road maintenance and construction. But in addition to its new mission of mass transportation oversight, the legislature that same year awarded the department the management of Texas' segment of the Gulf Intercoastal Waterway, in partnership with the US Army Corps of Engineers. Suddenly, what had always been a wheel-and-axle-, road-and-bridge-centered organization was now expected to designate dumpsites for

A new name and a new emblem.
(TxDOT Photo)

sand-dredging operations conducted in salt water, help clean up oil spills, and in years to come, even wind up running the ferry systems for Port Aransas and Galveston–Port Bolivar on the Gulf Coast.

Eventually it would also branch out into developing the Aviation Division, designed to help "cities and counties to obtain and disburse federal and state funds for reliever and general aviation airports included in the 300-airport Texas Airport System Plan," which "identifies airports and heliports that perform an essential role in the economic and social development of Texas." The aviation function, it is true, would not be folded into the organization for another sixteen years. But the future already loomed: one day the former Texas Highway Department was destined to become the all-inclusive State Department of Highways and Public Transportation, and that day would arrive sooner than the early Texas Good Roads pioneers ever might have envisioned.

The shift started occurring under Luther DeBerry's guidance. By the time he retired in 1980, after forty-three years of dedicated departmental activity, he and his wife had moved at least thirty-three times in obedience to Dewitt Greer's instructions, overseen changes far beyond his previous superiors' conjectures, and shepherded the department through some of its bitterest and most rigorous trials to date. It may have been pure realty

TxDOT operates free vehicle ferries at Port Aransas and Galveston Bay, taking motorists and their cars and trucks between Galveston and Port Bolivar. The ferries at Port Aransas are named for former department engineer/executive directors, and the Galveston/Bolivar ferries are named for former highway commissioners. Shown is the Robert C. Lanier Ferry operating in Galveston Bay in 2010. (Kevin Stillman/TxDOT)

Luther DeBerry at home in East Texas, 1999. He retired from the State Department of Highways and Public Transportation in 1980. (Geoff Appold/ TxDOT)

accident that his last Austin home turned out to sit immediately next door to the Greers' house, but it was also poetic irony: the best of leaders became the best of friends, and few neighbors could have shared a deeper bond of interests.

The Third Man

Marquis Goode succeeded to the engineer-director post in the middle of summer, 1980. The final Greer-groomed trainee to step into Dewitt Greer's shoes, he remained there for six years. His career span in the department matched the period of heavy postwar construction that led into the Age of the Interstate and reached its conclusion just as the final interstate segments were being completed. After graduating from Texas A&M on the GI Bill in 1947 (he had spent four war years in an engineering unit in Europe), Goode officially joined the Highway Department. Like several of his contemporaries, he chose it as his professional venue in the midst of several other, more lucrative offers from the oil and aviation industry; the stability, consistency, and excellent standards of both the work and the agency's structure held stronger appeal for him than a higher salary.

His first assignment, based in Dallas, was helping to transform a section of US Highway 75, the main artery from Dallas to Houston and the precursor to I-45 (on which Goode was also to work later) into a modern con-

*Marquis (Mark) Goode Jr., 1985.
(Geoff Appold/TxDOT)*

*Governor Mark White appointed
former Houston mayor Bob Lanier as
commission chairman. They led the
effort to increase the state gasoline tax twice, in 1984 and 1987, to
provide additional funding for highway
construction and education. (TxDOT
Photo)*

*November 1984 State Department of
Highways and Public Transportation
Commission meeting. Left to right:
Commissioners John R. Butler Jr.,
Robert C. Lanier, Chair Robert H.
Dedman, and Engineer-Director M. G.
Goode Jr. (Geoff Appold/TxDOT)*

crete four-lane highway. Eventually he was promoted to district engineer in Lufkin, overseeing the operations in nine counties. In 1973, he moved to headquarters in Austin to become the assistant engineer-director for Engineering Operations under his good friend and colleague Engineer-Director Luther DeBerry.

When DeBerry retired in 1980, Goode had no idea that he was considered the front-runner for the top position until the commission called him into a meeting, which of course included Dewitt Greer. "They didn't have a

One of the most exciting archaeological finds occurred in 1982 during excavation of a site near Leander prior to widening RR 1431 northwest of Austin. Highway Department archeologist Frank Weir led a team that had already found a rich trove of artifacts when on December 29, 1982, researcher Mike Davis found an intact bone, which eventually proved to be an intact skeleton, one of the earliest-known and most complete Paleo-Indian skeletons ever discovered in North America. Department photographer Geoff Appold dubbed the woman the "Leanderthal Lady." Carbon dating estimated she had been buried in this location ten thousand to thirteen thousand years before. (TxDOT Photo)

lot of conversation. They just asked me if I was interested in doing [it] and I said, 'Yes, I am.'"

Several initiatives occurred during Goode's service that would benefit department work and the state, including, in 1984, the first gas tax increase passed by the legislature in thirteen years. This increase had been aggressively sought by Governor Mark White's new commission chairman appointee Bob Lanier of Houston, a longtime transportation proponent. His fellow commissioners, previously appointed by Governor Bill Clements, were Bob Dedman and John Butler. (One of these had replaced longtime Clements supporter and Texas Turnpike Authority Commissioner Ray Barnhart, who had now moved on to Washington as President Reagan's pick to run the Federal Highway Administration—the second Texan to head the office, after Frank Turner—and who spent the next six years fighting to defend and preserve the federal road-aid funds and fuel taxes strictly for transportation.) Together the three commissioners overrode the no-rise policy long favored by Dewitt Greer and pushed for better-supported funding that then resulted in another rise two years later. By December 1985, contracts for highway construction were being let at the rate of $200 million per month.

Trash

Goode's tenure saw a new set of public information and public relations programs created in the state that literally had a worldwide impact—especially in conservation, environmental protection, and litter control. The first of these was the Adopt-a-Highway program, instigated by Tyler District Engineer James "Bobby" Evans and Public Information Officer Billy Black. For many years, the department had spent millions of dollars and expended thousands of hours trying to clean up highways, shoulders, and the landscapes edging them because the public annually demonstrated its indifference toward monitoring its own litter. Some of the worst offenders were careless young people. Bobby Evans was driving down a local highway one day when he saw, just ahead of him, a pickup truck spewing empty beer cans from the cab as the driver tried to lob them backward through the open side window into the truck bed. The cans inevitably bounced out of the bed or off the truck sides—if they got even that far—and onto the road. Traveling in the wake of this aluminum rain squall, Evans was struck—not by a flying beer container but by an inspired thought: If people could be convinced to actually take part in the cleanup efforts as volunteers, the action of doing so would in turn result in the limitation and/or considerable lowering of the amount of litter thrown out of vehicles in the first place. The more aware people grew, from cleaning up a section of roadway themselves, sweating under the hot summer sun or shivering in the cold, the less they might transgress. And the more pride they took in their clean highways, the

Litterbugs (five-year-old kindergarten children) sponsored by Austin Jaycees and Girl Scouts for Anti-litterbug Week, March 16–23, 1963. (Jack Lewis/ TxDOT)

harder they would work to keep them that way, especially if the worst offenders—Texas males thirty-four years old and younger—could be enticed to get involved.

Billy Black organized the program, enlisting local civic, church, and club groups to each sponsor a 1-mile or 2-mile stretch of highway and send crews of their members to pick up the garbage. Signs were erected at the onset of the adopted miles, naming the group responsible for their upkeep. The Tyler Civitan Club was the first to enroll in the program, and therefore the very first batch of citizens to set a precedent and start a wave that soon washed over the planet. Fraternities and sororities, Protestants, Catholics, Quakers, and atheists, strip joints, celebrities, cattle ranches, law firms, tattoo parlors, boutiques, bakeries, body shops, and Boy Scouts—all kinds of groups, individuals, and businesses soon joined the network. The signs proclaimed their participation, advertised their sense of duty, enhanced their public standing, and gave them all a sense of connection and shared purpose. The program proved breathtakingly fruitful. Shortly, other states copied its example, followed by other countries. Since its inception, forty-nine states have implemented Adopt-a-Highway plans, as well as Puerto Rico, Canada, New Zealand, Australia, and Japan.

Because any organization can volunteer its services to take charge of a length of highway, the widespread program has occasionally triggered conflict—particularly in areas where hate groups thrive. For instance, at one time the Ku Klux Klan adopted a portion of Interstate 55, just beyond St. Louis, Missouri. A public protest and repeated vandalism of the Klan's participant sign prompted controversy, but because of the Fifth Amendment

Tyler Civitan Club volunteers cleaning up their adopted highway section. (Kevin Stillman/ TxDOT)

Litter pickup by TxDOT volunteer during the Trash Off in April 2009. (Michael Amador/TxDOT)

of the Constitution, the program remained legally obligated to allow every group to join, without barring anyone. The State of Missouri approached the dilemma with a novel solution that also seized the opportunity to revere a noble civil rights pioneer: in November 2000, that same section of the interstate was designated as the Rosa Parks Freeway. As the KKK failed to fulfill its trash duties, its sponsorship was finally dropped. But other groups with axes to grind have also claimed their right to Adopt-a-Highway. Both the American Nazi Party and the National Socialist Movement, notorious for their white supremacist and anti-Semitic views, have assumed sponsorships in Oregon and Missouri. In the Missouri case, the highway section in question was again redesignated; it became the Rabbi Abraham Joshua Herschel Memorial Highway, named after a refugee who fled from the Nazi's advance in Europe and rose as a prominent theologian and civil rights advocate in the United States.

The second initiative toward trash consciousness launched under Mark Goode's aegis was the "Don't Mess with Texas" advertising program. Head of Travel Information Division J. Don Clark urged the department to contract a professional advertising firm to come up with a slogan and a series of television commercials that would discourage littering and exhort the public to clean up after themselves. The money spent on this campaign, he argued, would be more than made up for by reduced maintenance costs. GSD&M, an Austin-based company, took the job on, but it was not until one of the founders and partners, Tim McClure (who had grown up in Corsicana—the same hometown that produced the Highway Department's first state engineer, George Duren), remembered that during his boyhood his mother would declare his room a mess, that he conceived the warning cry that defined the crusade and became immortal: Don't Mess with Texas.

Branding the new anti-litter campaign. (TxDOT Photo)

The first bumper sticker emblazoned with that slogan appeared in December 1985. The first commercial aired on New Year's Day 1986, during the Cotton Bowl. It featured legendary guitarist Stevie Ray Vaughn sitting on a stool in front of a giant Texas flag, rendering "The Eyes of Texas" with true Hendrix-and-Vaughn blues virtuosity, while a voice-over announcer intoned, "Each year we spend over twenty million dollars picking up trash along our Texas highways. Messing with Texas isn't just an insult to our Lone Star State. It's a crime." Vaughn lifted his head, looked straight at the camera between the two final bars of his riff, and said, "Don't Mess with Texas." Immediately, the phone lines of local television stations jammed with callers pleading with the stations to repeat "that new Stevie Ray Vaughn music video about Texas." Almost all the callers were young males, the very offenders the commercial had been designed to reach.

Following that inaugural triumph, the department immediately commissioned future commercials, which starred a fleet of Texas icons, motorcycle gangs, cowboy poets, music masters, mistresses, and mavens, a World War II bomber, wrestlers, a baseball player, a Heavyweight Champion of the World (and Word—he is also an ordained minister), and even a monstrous

Stevie Ray Vaughan recorded the first "Don't Mess with Texas" TV spot at the Austin City Limits studio in January 1986. (Geoff Appold/TxDOT)

The Confederate Air Force with headquarters in Harlingen, now the Commemorative Air Force, filmed this dramatic spot on TX 349 in Upton County, March 1990. (Kevin Stillman/TxDOT)

"Mamas tell all your babies Don't Mess with Texas." Willie Nelson delivered his anti-litter message in this spot filmed on a Hill Country highway in June 1989. (Geoff Appold/TxDOT)

creature emerging from the Gulf surf and wheeling back under the waves after catching sight of the littered beach. Willie Nelson, Lyle Lovett, LeAnn Rimes, Joe Ely, Jerry Jeff Walker, Marcia Ball, Little Joe y la Familia, the Texas Tornados, the Fabulous Thunderbirds—all these and many more contributed their musical talents to the message, setting an example and alerting the target audience to their responsibilities. The commercials were catchy, clever, succinct, entertaining, and best of all, effective. Twenty-six of them screened over a twelve-year period, and by the time the campaign officially ended, the phrase they had made famous resounded across the nation, both as an emblem of Texas pride and an expression of Texan character. It was later updated by the Sherry Matthews Advocacy Marketing agency. The blanket incorporation of social media also helps send the message out.

That initial GSD&M outreach ushered in a new age of public-service announcements disseminated from other outside consultants hired by the department. EnviroMedia was the next firm chosen to carry on the "Don't Mess" mission. Safety promotions, including the famous "Click It or Ticket" seat belt campaign that the Sherry Matthews agency splayed across billboards, accompanied the Texas-based television ads and logos reflecting the ones the federal government had put to national use. Other Sherry Matthews collaborations also proved highly effective, such as the Christmas ads imploring people not to drink, drive, and wind up hitting Santa Claus, and the campaign featuring the victim of a drunk driver, a beautiful young woman named Jaqui Saburido, who was terribly burned and disfigured in an Austin car collision and who eventually appeared on the Oprah television show three times. This segment of the drunk driving campaign reached one billion people worldwide; it has since been instrumental in reducing drunk-driving fatalities in Texas by 20 percent and is widely regarded as perhaps the most powerful anti-drunk-driving publicity ever mounted in the United States.

Heroes

Through the last one hundred years, the staff of the Texas Highway Department has stood ready and committed to helping with crises over and beyond their construction and maintenance directives. In hurricanes and snowstorms, across floods and washed-out bridges, after tornadoes and explosions, the department's personnel have historically sped to the rescue with equipment and muscle power, handling emergencies, clearing evacuation paths, and assisting with the aftermath of disasters. For instance, the district engineer of Beaumont, John Barton, led the massive evacuation of Southeast Texas during Hurricane Rita in 2005, when the hurricane destroyed 4,526 single-family dwellings in Orange and Jefferson Counties alone, crushed or washed away mobile homes, and battered apartment complexes into ruins. Major damage was sustained by 14,256 additional single-family dwellings, while another 26,211 single-family dwellings also suffered damage. All in all, Texas reported 113 deaths—the highest number from that particular hurricane. Barton took on the responsibility to manage the recovery efforts—as he explained, "getting roads cleared and properties opened back up, helping emergency response teams get in, and then helping the community to restore itself over a series of several months. I was out of my home for seven months, I lived in my office, sleeping on a mattress on the floor, working twenty hours a day or more. It was a very humbling experience for me. . . . The teamwork that was created out of that experience, the bonding that occurred, is irreplaceable. I would never want to live through that again, but I wouldn't trade anything in the world for it."

But on occasion, the department's teams' or individuals' efforts rise above and beyond the basic calls of duty, when a few extraordinary heroes and heroines literally risk their own lives (and sometimes nearly lose them) in order to save someone else's. Their actions involve a member of the public—usually a motorist—and often occur during crashes, blizzards, twisters, or flash floods. Ambulances stalled in 5-foot West Texas snowdrifts with a critically ill patient trapped inside; motorists washed away in walls of water and rescued by a department worker willing to dive into the swirling rapids; a confused driver smashing his car into a ferry landing, staggering out amid ripped electrical cables, lurching straight into Galveston Bay, and paddling away from ferry staff manning the lifeboat to fish him out: these are a few of the scenarios in which department employees switch into high rescue mode. Prior to 1983, these remarkable people received a brief paragraph of description in the interagency newsletter, *Transportation News*. But that year, the department created a special award to honor them. Since then, the stories of their feats have been recounted at Short Course, in the same

Actions Above and Beyond

One example of circumstances that can prompt the award's issue is the case of four men from the Rio Grande Valley, all members of the State Department of Highways and Public Transportation road crew, who were working on US Highway 83 when they witnessed a rollover accident. To their horror, a sixteen-year-old boy was thrown from the vehicle during the rollover and flew over the median, landing on the highway where oncoming traffic was speeding his way. When one of the crew called for help on his two-way radio and then ran onto the roadway to yell and flag approaching drivers to the presence of the boy who lay injured on the street, his colleagues rushed to join him, and the four formed a human traffic block to protect the badly injured driver. Throughout this, the horrified boy remained conscious. Ultimately, thanks to their quick responses, he survived the accident despite his injuries.

In another incident outside Sulphur Springs, two department workers were returning to the local office after a day's routine when they witnessed a calamity: the driver of a propane tanker braked too hard to avoid hitting a cattle trailer, lost control, struck an oncoming vehicle, and then overturned. The men stopped, realized that propane was leaking from the tanker, and quickly took action to limit danger to others and assist the truck driver until other emergency responders could arrive. One began setting up traffic control and keeping additional motorists from approaching the scene. The other noticed the driver of the propane tanker was still inside and trapped. He ignored the eminent threat, and with the help of an arriving deputy sheriff, managed to get the driver out of the vehicle safely away from the dangerous area.

In yet another emergency, two employees in the Paris District were placing high-water barricades on roadways that had endured a large amount of rainfall and were starting to flood. When their work was complete, they drove up to a stranded vehicle with three women inside. One of the women began waving her arm out of the window, signaling for help. The two men knew they had to get the women out before the car was swept away. So they pulled alongside, once-a-year opening session in which the Gilchrist, Greer, DeBerry, and Stotzer Awards are presented to their recipients. The title of the award is "The Extra Mile."

Next in Line

When Mark Goode retired in 1986, his replacement had been working for the department as long as he had—specifically, since graduating from his same engineering class at Texas A&M in 1947. Raymond Stotzer, born and reared in Seguin, had served in the navy during World War II, then completed his degree and joined the department as a field engineer, assigned

lowered the "Tommy Lift" gate, and assisted the women in climbing out their car window to safety. In other similar but separate incidents, two mothers and their young daughters found rescue from TxDOT workers after their pickup truck and van had already filled with water and were on the verge of sailing down the churning flood, which they did just after the women and girls were extricated through the broken windows.

During the Christmas season, two maintenance technicians from the Fort Worth District who had been checking drains and clearing debris after recent rains were traveling along Interstate 20 when they noticed cars wildly swerving ahead and smoke billowing up from beneath the Clear Fork Trinity Bridge. By the time they reached the site, motorists were jumping out of their cars and running toward the bridge, and witnesses cried that a tractor-trailer driver had lost control of his vehicle, careened over the median, struck another vehicle, sending both vehicles over the edge of the bridge, and now they lay several feet below, between the east- and westbound bridges of the highway. One of the technicians ran to the bridge's edge and looked down to see the smoking eighteen-wheeler on its side on one of the frontage road turnarounds, with the wrecked car lying close by, and ran down the slope; the other turned his attention to the traffic that was backing up on the interstate, quickly set up traffic control to keep the other drivers safe, and cleared a path so the emergency response vehicles could reach the scene. Voices from beyond the crashed car called, "We got out!," assuring the first technician that the driver and passengers were safe. Meanwhile, the driver of the blazing eighteen-wheeler slumped behind his steering wheel, pinned in place by the crushed dashboard. A motorist climbed down the slope, and people from above passed down fire extinguishers and anything else they could find to stifle the flames, but the technician could not yet dislodge the driver. When another motorist and an off-duty firefighter scrambled down for a last effort, one of the men supported the trucker behind the dash while the TxDOT technician freed his legs. The technician then dragged him bodily out and carried him away from the cab just as the truck's fuel tank exploded, creating an inferno. The driver, the helpers, and the technician all lived.

Raymond E. Stotzer Jr.
(Jack Lewis/TxDOT)

to Guadalupe County. The versatile skills he used to impartially balance rural and urban needs and interests in Medina, Comal, Pharr, and Bexar Counties as he moved from one post to the next were duly noted by his bosses, and he was promoted repeatedly based on these merits. As he guided the Interstate 35 construction through Comal County, and then Interstate 10 through Guadalupe and Caldwell Counties, he also developed a reputation for excellence; when he oversaw the 1974 rebuilding of what was then the state's longest bridge—the 2.37-mile Queen Isabella Causeway in the Gulf connecting South Padre Island to the mainland—that reputation was confirmed. Like his predecessors, and like so many of his contemporaries whom the war had shaped through hardship and responsibility into practical, mature men who liked to tackle a job, get it done thoroughly and well, and take pride in the tangible results, Stotzer embodied thrift and high ethical principles.

It was at the beginning of his three years as department chief that a large number of those men and their old-school forerunners started retiring due to a new state-inaugurated incentive plan, and an entire generation of solid highway developers ended their migrations from district to district and slipped away into the landscapes of their hometowns, leaving the field open to the next—977 employees in all, with an average length of service of twenty-nine years. Stotzer's challenge was to reconstruct the department from a new generation of engineers and other personnel, with the same emphasis on team effort that had always been the department's strength. When asking the House Transportation Committee of the legislature for pay raises, he announced that although he felt the top departmental professionals deserved higher salaries, the lower-echelon employees should

Queen Isabella Causeway connecting South Padre Island to the mainland. (Geoff Appold/TxDOT)

Welder. (J. Griffis Smith/TxDOT)

receive them first; the department's team mechanisms could only be as strong as all of their working parts, and the rank and file deserved rewards sooner than their better-paid bosses. The legislature agreed and voted for pay raises for both groups, while the department praised Stotzer for his altruism and comprehensive vision.

Stotzer also turned into a fine recruiter, a magnet for talented and dedicated professionals who entered the department with exactly the kind of team attitude necessary to replace those who had left. One of the engineers he continued to promote to higher positions would replace him after his untimely death. Raymond Stotzer was diagnosed with cancer before his first three years of executive oversight had elapsed. He died on October 4, 1989, in Seguin, thirty-seven months almost to the day after taking the helm of the department. Less than ten days later, on Friday, October 13—one weekend before the annual Short Course Conference was to convene in College Station—his protégé Arnold Oliver took his place.

High Five Interchange at US 75 and I-635 in Dallas. (HNTB)

9

The Quiet Giant

The 1990s

Whatever job you have, concentrate on that and do the best you can.—Engineer-Director Arnold Oliver, 1989–1993

Engineers find ways to do for one dollar what any fool could do for ten dollars.—Deputy Executive Director John Barton, TxDOT, 2013–2015

To begin one's new job due to the rapid death of a mentor seems a sad, daunting prospect, no matter how capably one can meet its demands. To do so in the midst of preparations for a state legislature Sunset Review might amplify the stress just a little. To know that the future of one's entire agency hinges on that review's findings, and to successfully pass its close scrutiny only to then get ambushed by a state comptroller's audit, could perhaps even elevate one's blood pressure a degree or two. For the audit to end with

Wide-open spaces and wind warnings in West Texas, Culberson County. (Kevin Stillman/TxDOT)

an official command to immediately start cutbacks, staff reductions, and total restructuring might possibly quicken the pulse a little more. And to tackle all this at once, in a job one never sought in the first place, could just conceivably trigger daydreams of a long vacation in Tibet.

New Decade, Tumultuous Times

Much has been said in praise of the calm and stable objectivity of engineers. In a crisis there are few people more dependable; when it comes to problem solving, there is no profession more fearless and methodical. Arnold Oliver, an engineer to his bones, proved this assessment the moment he took the place of Engineer-Director Raymond Stotzer after Stotzer's brief four-month battle with cancer. He had not applied to fill the vacancy; when the highway commissioners informed him that he was their pick for Stotzer's replacement, he had been happily ensconced as Dallas's district engineer. His reply to Commission Chairman Bob Dedman's question as to whether he thought he could handle the task was typically self-effacing and realistic: "All I can promise is that I'll bring the same characteristics I bring to all jobs. I'll do the best I can."

Arnold Oliver was the first chief executive since Dewitt Greer's retirement not to be a veteran of World War II. A child of eleven when the war ended, he was already familiar with hard times and challenges. He had grown up poor, a Dust Bowl sharecropper's son who shared the life of migrant workers with his family in California during the mid-1940s before returning to his native Texas, where his father raised cotton and corn and struggled to clothe his family in pants and shirts made from cotton-sack canvas and flannel that Oliver's mother sewed at home. After attending

At an I-20 ribbon cutting in January 1998, Arnold Oliver addresses the audience while Commissioner Robert Dedman looks on (on the right, facing the camera). (Geoff Appold/ TxDOT)

Midwestern University in Wichita Falls for two years on a football scholarship, Oliver transferred to the University of Texas but found its tough academic standards hard to satisfy. He took a break to earn more college money, worked for the Highway Department for nine months, and there discovered his true calling: engineering. After a two-year stint in the US Army, he returned to the University of Texas, received his degree in 1960, and immediately applied for a full-time department job that moved him into the Wichita Falls District. There, among other assignments, he helped pioneer the uses of the new automation systems.

In 1972 Oliver became the resident engineer in Graham and a pioneer once again, this time in a different kind of organizational experiment. For many years, the gulf of mutual disdain between some of the Construction and Maintenance Divisions of the department had deepened, until hostilities between the two entities made cooperation thorny at best and easy communication almost nonexistent. Raymond Stotzer had addressed this problem while, as district engineer in San Antonio, he rebuilt the huge, convoluted Fratt Interstate that involved I-35, Loop 410, and major arterial streets in the northeastern part of the city by using lapsed interstate money from other states with no ready-to-go project plans. A district engineer, he reasoned, should be able to represent both halves of the department on a local level and present a unified front to the public. So when the maintenance supervisor in Young County retired, Bob Schleider, the Wichita Falls district engineer, considered it a rich opportunity to follow Stotzer's advice and heal the rift between the two operations by placing them both under the guidance of Graham's resident engineer, who would then become maintenance supervisor as well. That man was Arnold Oliver.

Eventually Raymond Stotzer moved Oliver again, placing him first back in Wichita Falls as acting district construction engineer, then in the top position at the Paris District one year later. From there he persuaded Oliver to take over the Dallas office, where Oliver engrossed himself in the planned rebuild of Central Expressway. But he remained there only two years. After Stotzer's death, Oliver suddenly found leadership conferred on him. He also inherited the cluster of ordeals that had been ripening during Stotzer's direction. And the first of these was the Sunset Review.

Since 1977, Texas has sustained a Sunset Advisory Commission, a legislature-created body intended to appraise and evaluate the effectiveness of any state agency—its management, use of public funds, methods and purpose, ongoing relevance—in short, its justification for continuing to exist. If an agency fails to meet the necessary standards on any of the criteria, the state government will simply choose not to authorize a funding renewal, and that agency will disappear. Much proof is required to submit to the reviewers; good records and transparency, as originally upheld in the precedents set by Gibb Gilchrist and Dewitt Greer, are essential, for the inspection is a thorough process—tantamount, as one high-ranking employee put it, to "an Inquisition." Through six special sessions, called after the regular session had ended—a record number, never since broken—department matters generated testimony and hectic debate. Fortunately, thanks to Arnold Oliver, Commission Chairman Bob Dedman, Commissioners Ray Stoker and Wayne Duddleston, and many others, the legislature finished its review by delivering only a few light blows as criticism and recommendations. The next blow, however, fell far more sharply: appropriately enough, its executor was State Comptroller John Sharpe.

Back during the administration of Mark Goode, the Del Rio District (District 22) had declined in utility. It was therefore closed in 1982, and two of its rural counties, Austin and Matagorda, transferred from the urban Houston District into the Yoakum District, complete with their own resident engineers and maintenance offices. Four of the other counties—Val Verde, Edwards, Real, and Kinney—were reassigned to the San Angelo District. The rest—Uvalde, Maverick, Zavala, and Dimmit—went to San Antonio. Del Rio's residency remained open, however, with its maintenance office and staff. But this earlier culling had been judged insufficient. Now the state comptroller issued his edict: in order to trim state expenses and help the budget, the State Department of Highways and Public Transportation would be required to whittle its number of districts from twenty-four to between eighteen and twelve and lay off about five hundred staff members. The comptroller's report became legislation in 1991. Its burden, of course, fell upon Arnold Oliver.

During this period, the agency had been expanding in other ways: the Environmental Affairs Division and the Civil Rights Division had already

North Central Expressway ground breaking included Oliver smashing a construction barrel. (TxDOT Photo)

Texas
Department
of Transportation

Another new name and a new logo.

been created, and the department added a statewide bicycle coordinator as head of the Bicycle Advisory Committee, a part of the Public Transportation Division. More were shortly on the way, when in September 1991, the name State Department of Highways and Public Transportation changed to the far simpler and more inclusive Texas Department of Transportation, or, as it is known today, TxDOT. The Motor Vehicle Commission (which one year later became the Motor Vehicle Division) joined the department's list, as did the former Department of Aviation, now known as the Aviation Division. Although the word "public" had been dropped from the name, actual civic participation enlarged through the newly implemented advisory committees. Another title change also came along with the fresh packaging: no longer would the head of the agency be referred to as the engineer-director. Arnold Oliver, and his successors, would henceforth be known as the executive director—a name that later carried a weightier implication, now that the historic "engineer" label had vanished.

Meanwhile, Oliver, his task force, and all the employees hovering with bated breath to see what their fates would be, were coping with the details of reorganization and their diving spirits as quickly and efficiently as possible, working against the deadline pronounced by the legislature. Tensions tightened. The Texas Transportation Commission had studied the newly suggested district boundaries and approved them in early November: in alphabetical order, Atlanta, Brownwood, Childress, Lufkin, Paris, Waco, and Yoakum were slated for downsizing, another word for the executioner's blade, while a new district based at Laredo took shape on the map.

But other elements entered the equation and were soon factored into the final decisions. The legislature raised the gas tax to 20 cents per gallon —a nickel more than the previous increase five years before and an amount that, as of the spring of 2015, has remained fixed in place for twenty-four years. Commissioner Ray Stoker, who had worked so strenuously

The state Transportation Department provides infrastructure connectivity along the border with Mexico. Laredo, located at the southern end of I-35, is the busiest commercial land port in the United States. The first bridge connecting Laredo and Mexico's Nuevo Laredo was built in 1899. Today, there are four pedestrian and vehicle bridges and one rail bridge in the city. A fifth commercial bridge is in the planning stage. (Michael Amador/TxDOT)

for "nearly twenty-four hours a day," according to his executive assistant Maribel Chavez, to get the tax raised to 25 cents per gallon that he had a heart attack, was doomed to disappointment when, during his cardiac surgery, the legislature raised the tax by only half that amount (the department benefited from only 3.5 cents of the nickel, as the Department of Mental Health and Mental Retardation and the Department of Public Safety "borrowed" the remainder from the highway funds). Also, just at the end of that same momentous year of 1991, President George H. W. Bush signed the new federal transportation bill, the Intermodal Surface Transportation Efficiency Act (which, among other objectives, defined a number of High Priority Corridors to be part of the national highway system), that increased federal funding to Texas by about $385 million a year. By February 1992, the House had halted TxDOT's remit to reorganize, suspending all plans and efforts toward that goal until at least June 1993. The ultimate outcome: instead of the number of districts pared down to twelve, with Laredo's addition, it climbed back up to twenty-five (the original six districts designated in 1917 are in italics): Abilene, *Amarillo*, Atlanta, Austin, Beaumont, Brownwood, Bryan, Childress, Corpus Christi, *Dallas*, El Paso, *Fort Worth*, *Houston*, Laredo, Lubbock, Lufkin, Odessa, Paris, Pharr, *San Angelo*, *San Antonio*, Tyler, Waco, Wichita Falls, and Yoakum.

But the labors of Arnold Oliver did not cease there. Because his ascendancy to the executive post occurred only one year before Ann Richards was elected to the governor's chair, Oliver soon had another new problem with which to wrestle: the pressure to diversify the department by hiring more minorities and women. As the first female governor elected in Texas since Miriam Ferguson, Richards made it clear that one of the foremost aims she pledged to her constituents was the institution of a more balanced state personnel. "Today, we have a vision of a Texas where opportunity knows no race, no gender, no color—a glimpse of the possibilities that can be when the barriers fall and the doors of government swing open. Tomorrow, we have to build that Texas," she announced in her inaugural address on January 8, 1991. And TxDOT, famously the domain of white males (traditionally the social category most likely to produce civil engineers, maintenance recruits, and construction workers), needed to comply.

A lawsuit filed in the early 1980s by the US Justice Department had first issued the charge that the State Department of Highways and Public Transportation was guilty of discrimination in its employment practices. Other state agencies had also been accused at the same time but had quickly settled by turning over the responsibility of internal personnel matters to the government. But after investigating the various complaints filed by disgruntled witnesses, the department determined that only three "had even dubious validity," according to Engineer-Director Mark Goode; therefore, the department decided to call their bluff and ride the suit all the way to court. Many subpoenas were issued to district engineers, managers, and administrators; much paperwork shuttled back and forth between Austin and elsewhere to Pecos, where the trial was held. After four days, the federal judge sitting on the bench closed down the testimony; about a year later, in the fall of 1982, he dismissed all charges and ruled that the Justice Department should reimburse the state for all its costs. The State Department of Highways and Public Transportation stood exonerated—for the moment. As Mark Goode said, "The only question we asked is, 'Are you qualified for the work?' All people, including minorities and women, achieved higher levels as a result."

Already, the first African American engineer had been employed by the agency since gaining his engineering degree from Prairie View A&M University in 1969: Walter O. Crook, who eventually retired in 2003 after thirty-five years of service. In a number of areas, the old customs and expectations had gradually been thawing; a good example of this trend was Maribel Jaso Chavez, who in 1989 had become the first female resident engineer, based in her hometown of Pecos. In 1991 she was appointed executive assistant to Commissioner Ray Stoker, and in 1992, as the first woman, the first Hispanic woman, and the youngest person in the department to be promoted to district engineer during that period, she took on the Abilene District—in

Beaumont District Engineer Walter Crook. (Michael Amador/ TxDOT)

Ground breaking for the new Denison Travel Information Center in August 1993. Governor Ann Richards and Commissioner Anne Wynne, along with local dignitaries, wield shovels. (Geoff Appold/TxDOT)

more than one sense. Eventually she moved to the district engineer position in Fort Worth. There had long been many women who worked in the various divisions, as well as in the district offices. But the deficit of visible minorities and women in the agency continued to chafe critics, and soon TxDOT had a roster of new hires and administrators, including the first Hispanic commissioner, Henry Munoz III, and, in 1993, the first female commissioner, Anne S. Wynne.

Maribel Jaso (Chavez) and Cora Sue Harrison. "She was the first female maintenance supervisor at the department," recalls Chavez. "She was named just a month or so before I was as resident engineer at the same office at Pecos. The Pecos residency oversaw Reeves, Ward, Loving, and Winkler Counties (in West Texas). She had Reeves and Loving for maintenance. This is right before Mr. Stotzer combined maintenance and engineering in the field. Who knew that West Texas was so progressive?" (Courtesy of Jodi Hodges/TxDOT)

The Surge of Change

As pioneers so often discover, breaching a frontier is not always pleasant. "I put myself through school working in the oil fields," explained Maribel Chavez, the first female district engineer in departmental history. "I worked for Gulf Oil in the summers as a roustabout. That's what you do in the Permian Basin. Two of my older brothers are engineers, one chemical, one petroleum. . . . When I was in high school and they both asked me what I was going to be, I said, 'I don't know,' and they said, 'You're going to be an engineer.' 'I am?' I said. 'Yeah, you are. What kind do you want? Pick one.' When I was little I used to watch the television show *Family Affair*, with [the character] Uncle Bill who went all over the world building bridges. Fascinating. So I said, 'I guess I'll be like Uncle Bill.' 'Okay,' said my brothers. 'That's it. Settled.'" A former homecoming queen in her Pecos high school, Chavez completed her civil engineering degree at the University of Texas. Then, hired by the State Department of Highways and Public Transportation on the same day she submitted her application, Chavez found out what it meant to be part of the "Highway Department family." Within one month of starting her new job, she watched her father suffer a major heart attack. The instant her colleagues learned of it at work the next day, they sent her straight home for the duration of the crisis.

The gender politics generated by her arrival presented several conundrums, including the problem of what to do with her, the sole female from her district eligible to attend the week-long Short Course alongside a carload of males. This was resolved when she got permission to invite another woman along—a solution that, incidentally, fulfilled the other employee's long-time desire: Lydia Jacobs, draftsperson in the Odessa District design office, had waited years to finally be invited to a Short Course. But there were other, less happy moments. Chavez's promotion from Odessa to the resident engineer's headquarters at Pecos placed her one block away from her childhood home, which also created controversy; during her presentation of an unpopular proposal to the Pecos City Council, a commissioner who had known her since her elementary schooldays warned, "Maribel, if I have to, I'm going to call your parents."

A little less than a year later, she received the summons from Highway Commissioner Ray Stoker to become his executive assistant. After serving his administration for two years, she resumed engineering, and at Executive Director Arnold Oliver's urging, she applied for the position of district engineer. Three positions stood open due to the recent retirements sweep: Childress, Yoakum, and Abilene. Erroneously perceived as Governor Ann Richards's "token female, token Hispanic" selection, Chavez ran the roughest gauntlet of her life when she accepted the Abilene post. The engineer over whom she would be presiding, who had already "cussed out" Arnold Oliver for her appointment, told her during her introductory phone call, "You and I need to talk. But you cannot come to the District Office yet." Instead, he suggested they meet at a neutral spot some distance from Abilene. When Chavez arrived at the McDonald's in Big Spring, the man curtly greeted her and then, as Chavez said, "proceeded to read me the riot act. He said, 'Let me just be straight with you. Nobody wants you here. We all think the only reason you got selected was because of Ann Richards. We don't think you deserve the position, and most of the staff is not happy. They wanted me to be the District Engineer. I'm not going to make trouble for you, but I don't know if people want to work for you.'" Chavez realized that he and she were both caught in the middle of a huge culture shift. She replied, with deference to his seniority, that she was now the DE, and they would just have to see how things proceeded.

"The next day I went to the District Office. Needless to say, I was not welcomed. The DE secretary had been there for many years, through many district engineers, and I could tell she was really nervous, almost like she was embarrassed, like I was about to witness some

Maribel Chavez.
(Geoff Appold/TxDOT)

really crazy stuff. She said, 'Well, Maribel, I'll take you to your office.' I walked in [to the office] and it was empty, except for an overturned trash can in the middle of the room. There was no furniture. I could tell [the secretary] was so uncomfortable—bless her heart, *I* felt sorry for *her*. So I said, 'Well, I get to remodel. I get to bring my own furniture.' I think it was a relief for her. I thought, 'Well, here we go. Welcome to Abilene.'

Chavez then asked her secretary to call a staff meeting so she could introduce herself. When she walked in, she saw that "it was all men, and they were all older—a whole lot older. The youngest, least tenured of the group, Design Engineer Billy Jackson, had only been working for the department for thirty-one years. The construction engineer had thirty-eight, the maintenance engineer forty-two. They didn't have two words to say to me. They were not happy campers." She remained for another seven years; meanwhile, the unfriendly veterans quickly took advantage of the new retirement incentives. "Literally, in one day," Chavez said, "I had fifty-four people retire on me. The entire staff—every single district supervisor, all thirteen maintenance supervisors. Everybody in the lab: six people. From the warehouse supervisor to the equipment supervisor. All of them. Three area engineers left; only one stayed. Back in design, we cleaned house. There was nobody there, nobody. For one month, all I did was interview and hire."

Chavez developed a brand-new workforce from scratch, all of them younger and less experienced than she and several of whom went on to forge illustrious achievement records in the department. Eventually, following Wes Heald's promotion to TxDOT's executive director-ship, Chavez became district engineer for the Fort Worth office and received a Preservation Leadership Award for her efforts in preserving the community's historic bridges. "Her inclusive management style fostered honest conversations that hold promise for the future," Jerre Tracy, Historic Fort Worth executive director, said of Chavez. "This type of respectful dialogue be-tween a state agency and non-profit organization creates opportunities for Fort Worth's future, and no one does it better than Maribel Chavez."

When Chavez retired from TxDOT in 2013 after twelve years in the Fort Worth district engineer position, her successor, Brian Barth, proved to be an ideal amalgam of both old and new department traditions and alliances. Also a University of Texas engineering graduate, Barth grew up in El Paso, working throughout his high school and college summers construct-ing Texas Highway Department projects for his father, Richard Barth, cofounder of the giant contracting firm J. D. Abrams. The senior Barth naturally hoped for Brian to follow in the family footsteps. But Brian Barth developed other ideas. "I decided I wanted to be the guy designing those plans," he said, "rather than the guy doing the take-offs and figuring out how to build them." After graduation, he first became engineering assistant in the Dallas District, then the Dallas District's director of transportation planning and development, and on to the Fort Worth District's deputy district engineer, to overseeing a statewide initiative in 2013 to streamline the procurement of professional engineering services to help TxDOT deliver the most transparent and cost-effective projects, culminating in his current post—the classic curve of a modern-day DE's career.

Arnold Oliver's stint in the chief executive chair also saw certain environmentally slanted advances take place, one of which was the application of recycled asphalt pavement (RAP) on road surfaces. The first project in which RAP was put to use was I-35E, just south of Dallas—a 100 percent recycled materials undertaking that foreshadowed the future. In the following years, its processing was refined, with formulas that now balance flexibility, sturdiness, and maximum reuse; alongside it, the development of recycled asphalt shingles (RAS) enlarges the available construction resources without depleting the earth of virgin asphalt stocks (crude bitumen, or pitch) as severely as in the past. Other formula sets of asphalt binders and hot mix, called Superpave (Superior performing asphalt pavement), also improved paving durability. Water-based paints now striped the highway. Rubber was recycled according to new federal mandates. The vegetation management standards changed, with an emphasis on curbing the amount of mowing done on rights-of-way. Advanced Traffic Management Systems involving new telecommunication connections (which would come under the umbrella of Intelligent Transportation Systems) were installed in urban areas, with commanding walls filled with multiple screens with which to observe live traffic on any given major freeway artery. The metropolitan districts of Houston, Austin, Fort Worth, El Paso, San Antonio, and Dallas all established stations with these up-to-date facilities.

Sealcoat operations in the San Antonio District. (Michael Amador/TxDOT)

Under modernized vegetation management practices, native grasses and other native plants thrive with little help from highway maintenance crews. (Stan Williams/TxDOT)

The landscape explodes with natural beauty on the River Road (FM 170) from Study Butte to Presidio through Big Bend Ranch State Park. (Kevin Vandivier/TxDOT)

After Oliver: Bill Burnett, 1993–1997

In 1993, William G. "Bill" Burnett, who had been the district engineer for Abilene immediately before Maribel Chavez, took the place of the retiring Arnold Oliver as executive director of TxDOT. His engineering and administration credentials made him an excellent candidate, in the continuing tradition of the Department. Although the only top executive to have been born outside the state since Dewitt Greer (whose Louisiana birth could be called a momentary detour, since his Texan parents stayed there only briefly before returning to East Texas when Greer was three weeks old), Burnett

came to regard Texas as his home under odd circumstances. Born in Engle-
wood, New Jersey, in 1947, he moved with his family to Toledo, Ohio, as a
small child and from there to Tucson, Arizona. In 1960 Burnett's father, an
FBI agent, relocated the family to his newest posting in Puerto Rico. Then,
in 1963, the FBI transferred Agent Richard Burnett to Dallas to investigate
the assassination of Present John F. Kennedy. After the Warren Commission
concluded its report, the bureau reassigned Agent Burnett to head its San
Angelo area office—just in time for Bill Burnett to enroll in San Angelo High
School for his senior year. Thus, the tragic murder of a US president transformed a native New Jerseyan into a Texan.

Burnett graduated from Texas Tech University with a civil engineering
degree in 1971 (the only engineer executive director not to have attended
either the University of Texas or Texas A&M University). After having worked
for the Texas Highway Department as an engineering technician in Ozona
during two of his college summers, he joined it as a full-time employee
upon graduation. In 1986, Burnett became the second-youngest district
engineer the department ever hired (Dewitt Greer was the youngest), when
he was promoted to lead the Abilene District. Five years later, in 1991, the
department named him district engineer of El Paso.

This career trajectory beautifully demonstrates the professional arc described by Marcus Yancey, when an engineer of true merit, dedication, skill,
and service rises "like cream" to the top of his calling. In 1993, the Texas
Transportation Commission unanimously chose Bill Burnett as executive
director of the Texas Department of Transportation. He was, at forty-four,
the second-youngest director the department had ever seen (again, only
Dewitt Greer, thirty-eight at the time of his appointment, had been younger). He also, in 1996, followed in Greer's footsteps to become the sixth Texan
to fill the presidency of AASHTO.

William G. "Bill" Burnett, 1994.
(Geoff Appold/TxDOT)

Congress Avenue Bridge, Austin

When the Congress Avenue Bridge was opened on April 4, 1910, it was engineered to withstand frequent Colorado River flooding that wreaked havoc through the middle of the capital city. The iron bridge pictured next to the new bridge was dismantled and removed. (HNTB)

Seventy years later, when the bridge was in need of replacement or expansion, engineering studies showed that the original supports were still sound, and designers had only to redesign the bridge decking, adding wider sidewalks and additional lanes. The newly renovated bridge was completed in 1980. The Austin City Council renamed the bridge after former governor Ann Richards in 1995. (John Suhrstedt/ TxDOT)

Upon completion of the bridge, Mexican free-tailed bats found that the gaps beneath the bridge deck were a perfect habitat. The bridge is now home to the largest urban bat colony in North America. In the summertime, thousands of people line the bridge and banks of Lady Bird Lake at dusk to watch up to 1.5 million bats emerge for their nightly foraging for mosquitoes and other insects. TxDOT instituted a program called "Bats and Bridges" to create designs for other suitable structures to provide habitat for bats. (J. Griffis Smith/TxDOT)

One of the most significant TxDOT achievements to occur during Bill Burnett's tenure as executive director was the wider implementation of the new Intelligent Transportation Systems. These eventually encompassed a number of functions to aid vehicle navigation through all kinds of conditions, such as controlled traffic signals, variable message signs, speed cameras, and the integration of live data culled from various sources: drivers' cell phones, weather reports, congestion reports and estimates, bridge deicing—any information that could immediately contribute to better traffic flow and mobility management and keep drivers alert to whatever might slow, impede, or endanger them a little farther up the road. By now, the term "intermodal transportation" (all the types of transportation for which TxDOT is responsible, including aspects of aviation, ferries, bicycle routes, passenger and freight rail, bridges, and of course, highways) had also come into full usage—a favored buzzword of the mid-1990s. And it was during this period that TxDOT ventured onto a new highway: the "information superhighway," as the Internet was then dubbed.

TransGuide, an Intelligent Transportation System, was designed by the TxDOT San Antonio District. This "smart highway" project provides information to motorists about traffic conditions, such as accidents, congestion, and construction. With the use of cameras, message signs, and fiber optics, TransGuide can detect travel times and provide that information to motorists not only with the message signs on the highways but also with the use of the Internet and a low-power television station. Similar systems in Texas' other large cities partner with local and county authorities to manage urban highways. (Texas Transportation Institute)

Traffic camera, which is part of the El Paso traffic management system. (Texas Transportation Institute)

The Sugarland Airport, southwest of Houston, is one of about three hundred Texas airports open to the public. The state's general aviation airport system is one of the largest in the nation. TxDOT's Aviation Division assists cities and counties in applying for, receiving, and distributing federal and state funds for reliever and general aviation airports. (TxDOT Photo)

Maintenance is an ongoing and required task, and painting the steel bridges is vital to keeping the structures sound. These department employees, Virgil Ingram (left) and Harold Ringer (right) are on the paint crew for the Rainbow Bridge. (Doug Fairchild/TxDOT)

Burnett's perspective of the TxDOT family—its astounding loyalties, the sense of mutual cooperation and public service unifying its members, the transcendence of any cultural differences, political biases, or personal animosities in regard to pitching together to get their jobs done—was archetypal. While attending regional meetings as a resident engineer with a young family in tow, he was touched and impressed when Engineer-Director Marquis Goode's wife, Lucille, immediately took them under her wing and saw to their comfort. As a neophyte in the Odessa office, he witnessed the protective care demonstrated by the district's secretary. Before the invention of direct deposit into banks, department salaries were paid at the end of each month through a "Pony Express" system of relays, whereby an employee would leave agency headquarters in Austin with a satchel full of checks, move westward (or in other directions radiating outward from Austin) to the next district, where he would drop off that district's checks and hand over the satchel to the next messenger in line, who would then continue the journey until, courier by courier, the paychecks were distributed to their proper owners. Sometimes the process would slow or encounter obstacles. Sometimes as many as ten days would elapse before the employees in a far district saw their earnings—and by then, the secretary would already be telephoning the local electric and water utility departments, explaining, "Please don't cut our people's services off. They're still waiting to get paid."

But Burnett had no illusions about his importance in the long-range picture of the department's continuity. As he once commented to a commissioner, "David, I want to tell you right now: there are 14,726 employees in our department. Of those, what the commission does and what I do, only 200 give a damn, because we really don't affect their daily lives. They just

Contract workers on a job in San Antonio. (Geoff Appold/TxDOT)

go out every day and do their jobs. If it's patching potholes, they patch potholes."

Before Bill Burnett left the department, he, as had his predecessors, saw it pass through the Sunset Review process in 1997 and receive the legislature's sanction to continue its existence and its work. When he retired after twenty-six years of service to Texas and the department, he stated, "There is no finer place to develop and grow as an engineer than TxDOT." He then entered the private sector, leaving the department that had nurtured him, and that he had in turn nurtured, in the hands of another, even longer-term veteran: the brilliantly solid Mike Behrens.

Hot-mix application on TX 46 near Boerne. (Kevin Stillman/TxDOT)

The newly constructed state highway building seen from the top of the State Capitol. (Robert M. Stene/TxDOT)

The Dewitt C. Greer Building, Then and Now

The Texas Highway Department Building, located at 125 E. 11th Street in Austin, directly across the street from the State Capitol grounds, had remained largely unchanged since its completion in 1933 when, over sixty years later, the decision was made to renovate and update it. The art deco facade and interior lobby had by then acquired historical as well as decorative significance; the addition of air conditioning in 1951 had been one of the few concessions to modernity enjoyed by the workers beyond that lobby. Built at the site of the old Travis County Jail on ground cleared of the jail's demolition debris by Highway Patrol cadets in training at Camp Mabry, the structure designed by San Antonio architect Carlton Adams originally cost $455,154.74. It features carved limestone finials on the two front columns, a carved limestone panel above the door, and matching panels that serrate the roofline. Inside the inlaid marble lobby is a bronze bas-relief frieze depicting transportation throughout the ages in Texas.

In 1981, the legislature changed the Texas Highway Department Building's name to honor the retiring former state engineer and commissioner Dewitt Greer. In 1995, strategies to pre-serve the integrity of the commission hearing rooms and offices while bringing them into the twenty-first century were overseen by the Texas Historic Commission. In 1998, the building was added to the National Register of Historic Places.

Entrance lobby of the Greer Building. (J. Griffis Smith, TxDOT)

The block at Congress Avenue and 11th Street in Austin contained the Travis County Jail (with turrets), county courthouse, and temporary State Capitol in the 1890s. This early photo is looking west from Brazos Street. (Courtesy of Texas State Library and Archives Commission)

The Travis County Jail was located at 11th and Brazos Streets along with the county courthouse. The building was replaced with the State Highway Building in 1933. (TxDOT Photo)

Art deco features embellish the Highway
Building's exterior. (TxDOT Photo)

(TxDOT Photo)

Ghosts have been said to haunt the Greer Building over the years. Occasional sightings, characterized by a peripheral glimpse of a figure prowling behind someone in a hallway and then slipping into thin air, or a schoolboy strolling past the night watchman's desk at 2:00 a.m., or a man dressed as a judge in the conference room (who also vanishes), have been reported by employees, sometimes with alarm, sometimes not. Unseen phenomena supplement some of the effects: file drawer handles rattling of their own accord, doors opening or closing without human touch, footsteps in empty rooms, coffee cups found turned upside down. One night a contract security guard performing required building checks during the midnight shift heard a door slam about 4:00 a.m. When she began to investigate, a loud, gruff voice commanded, "Go away." She did—right out of the building and onto the street, where she phoned TxDOT's security staff to announce that she was quitting. Other occupants staying late to answer road condition calls or tend to other tasks have felt a distinct aura of annoyance emanating toward them from some presence in their close vicinity. The fact that the types of personalities who work in the Greer Building (and Transportation in general) are usually by nature highly rational, logical, and business-like makes these experiences all the more startling, not to mention intriguing. Some attribute the disturbances to the old battlemented jail's history—in particular, the hangings that took place there through the decades, when counties were still responsible for local executions, before the electric chair was instituted in Huntsville.

Wes Heald, 1998–2001

Charles Wesley "Wes" Heald, a native Texan born in Spur in 1937, had already worked in the department for thirty-seven years when he became executive director. The son of a road construction worker, he left Spur very early and lived with his family in more than two dozen places before they settled on a ranch in rural Parmer County when he was five. He moved from West Texas to Evant, in Coryell County, where his father operated ranches, for junior high and high school. Then, like so many permanent careerists in the Texas Highway Department before him, he started by working on a road construction project during the summer of his junior year. That early introduction led him not only to an engineering future—a desire he already cherished—but specifically to aspire for a department job.

In 1960, Heald graduated from Texas A&M University with a civil engineering degree and signed on with a private contractor before entering the US Army that same year as a second lieutenant (he retired from the Army Reserves in 1984). After his discharge from active duty the next year, he worked for a short time in Houston's urban expressway office and then moved to Brownwood, where he joined the Texas Highway Department. His work maintained such a high standard that in 1970, the Abilene Chapter of the Texas Society of Professional Engineers named him Young Engineer of the Year. He became the Brownwood district engineer in 1987, the Fort Worth district engineer in 1993, and executive director in 1998.

Charles Wesley "Wes" Heald posing for a management team photo on the Greer Building steps, 2001. (Geoff Appold/ TxDOT)

Mary Lou Ralls served as state bridge engineer and director of the bridge division from 1999 to 2004. During her TxDOT tenure she led the development of one of the most attractive and efficient new precast bridge systems, the Texas U-Beam, and was in the forefront of the development and implementation of modern technology such as high-performance concrete bridges and totally prefabricated bridge systems. (Geoff Appold/TxDOT)

His restoration of the Bridge Division into a sole entity once more (it had earlier under Bill Burnett's decision been combined with the Design Division) also revived a sense of resolute pride and focused skills for design, construction, and renovation—qualities Heald felt to be waning in the department, as more projects were parceled out to private contractors. There are 52,260 bridges in Texas, the state with the most bridges in the United States (next is Ohio, with 27,901). Two-thirds of these come under the auspices of TxDOT. To resume complete responsibility for this system was a gratifying move, and Heald reassigned a number of Design and Construction Division employees to uphold the task.

On December 21, 2000, an event with far-reaching consequences and enormous implications for Texas and the nation took place in Austin: Governor George W. Bush turned over the Governor's Mansion to Lieutenant Governor James Richard "Rick" Perry, in order to free himself to assume the presidency of the United States. This moment marked a new epoch for the state and all its agencies and, unbeknown to those quietly toiling away among its corridors and on its roadways, signaled what would become a volcanic cataclysm for TxDOT. Following his official election to the governorship in 2002, Perry, who had graduated from Texas A&M Univer-

Former executive directors attend the dedication of the Wes Heald Ferry in 2011. Left to right: Wes Heald, Mike Behrens, Arnold Oliver, and Bill Burnett. (Randall Maxwell/ TxDOT)

sity with a degree in animal science, would continue in that office through the next twelve years, being reelected for two more terms to end as the longest-serving and therefore the most powerful governor in state history. This power hinged on his right to select or appoint every possible governor's choice of seat, judge's bench, agency head, officer, public university regent, and any other position dependent on his personal discretion—and he exercised that right for fourteen years.

When Wes Heald retired in 2001, he had accumulated forty years' service to the department and the State of Texas. At the Transportation Commission meeting recognizing his retirement, new Perry-appointed Transportation Commissioner Ric Williamson said of him (attributing these hymn lyrics to scripture in his remarks), "The department was once lost and it became found because Wes Heald took us back to the basics of perfection of building highways and running the transportation system, and as a taxpayer, I'm grateful." Considering Williamson's dream for the department's destiny, these words held highly spiced irony. Heald's successor, Mike Behrens, a thirty-one-year department veteran, previously the district engineer for Yoakum and the assistant executive director for Engineering Operations since 1998, was a strong and obvious choice to run the department.

And with what was coming, he would need to be.

Stars and arches are decorative elements on I-10 and Loop 375 in El Paso. (Kevin Stillman/TxDOT)

From 2001 to the Centennial

Every road is a toll road. It just depends on where you collect it.
—Dewitt Greer

If you add up the total number of lane miles in Texas, you can circle the earth 3.2 times.—Walter McCullough, retired district engineer

On September 15, 2001, only four days after the 9/11 attacks and only two weeks after Wes Heald's departure and Mike Behrens' ascension to the executive director's office, tragedy literally struck in Texas when a tugboat pushing four steel-laden barges and underpowered against the current rammed into the Queen Isabella Causeway that connects South Padre Island to Port Isabel on the mainland.

Construction repairs to the Queen Isabella Causeway in 2001 after a barge ran into and wrecked part of the bridge. (Kevin Stillman/ TxDOT)

Two 80-foot sections of the 2.37-mile bridge collapsed immediately, sending five vehicles with unsuspecting drivers hurtling 85 feet down into the sea just as they crested the rise of the bridge that now, invisibly to oncoming cars, yawned over empty space. Because the accident occurred at 1:30 in the morning, traffic was comparatively light. Nonetheless, eight people died. Four more were rescued by fishermen who witnessed the catastrophe. Fourteen hours later, a third section collapsed, leaving in total a 240-foot gap in the twenty-seven-year-old causeway that effectively stranded everyone still on the island in place.

This was not the first time the Queen Isabella Causeway had suffered from an operator's error. In 1996, the pilot of a small airplane had flown his Cessna TR182 under the bridge, then turned 180 degrees to maneuver another pass, at which time his plane collided with a concrete pylon and column, exploded, and dropped into the Laguna Madre in a fiery black plume of smoke. On that occasion the only casualty was the pilot. But the 2001 disaster not only cost innocent lives (although that of course was the worst aspect of the collision); the cost of the damage escalated into the multi-millions, included the island's electrical lines and freshwater supplies, and stirred anxieties (quickly allayed) that the crash was another terrorist assault.

By 9:30 a.m., TxDOT bridge designers were performing aerial inspection from a plane. By lunchtime, divers were checking the footings, while Pharr District Engineer Amadeo Saenz and representatives of Williams Brothers Construction assessed and sought answers to the problem. They quickly determined that the third section would probably fail, a prediction shortly confirmed when it tumbled down. In Austin, engineers with the Bridge Division reviewed the causeway's original construction drawings and materials listings.

The islanders remained trapped for nearly four days, after which TxDOT ferries from Port Aransas joined a collection of private vessels carrying people back and forth between makeshift docks, shortly augmented by another ferry loaned by North Carolina's DOT, as well as a privately contracted ferry from Mobile, Alabama. This temporary solution shuttled seven thousand vehicles and twenty-five thousand passengers per week for the next forty-five days, at which time TxDOT and its contracting partner Williams Brothers were able to complete the round-the-clock demolition and reconstruction of what became 400 feet of bridge (including the two sections on either end of the damage) thirty days ahead of schedule. "From my perspective," said South Padre Island Chamber of Commerce President Roxanne Harris, "all of the island businesses and residents remain eternally grateful for the rapid reopening of the causeway, not only for the immediate effect but also for the long-term impact."

Mending a Disaster

For the people marooned on South Padre Island after the Queen Isabella Causeway disaster on September 15, 2011, the momentary privations might have seemed trivial compared to their predicament. Any chance of reaching the mainland was limited to small, private ocean craft, a measure that precluded most road-based mobility once passengers set foot on shore, and although docks were in place that could serve people ferries, on the mainland, only a minimal structure existed that might support a car ferry dock—two piers left over from the old, original causeway built many years before the damaged one—and on the island, there was nothing at all. So Amadeo Saenz and his Pharr District crew of mechanics, welders, and shop workers set about inventing an emergency facility that could provide car access to ferries.

The first step involved building a ramp and anchoring it onto the mainland piers, with a mechanism to raise and lower it. "We designed it on the fly," Saenz said. "The first ramp we put in place—we had some of these tilt trailers that had a drop-down ramp, and I made them cut that ramp. The guy that drove the truck is probably still mad at me. 'That's a brand-new trailer,' he said. 'I understand,' I said. 'So we'll get you a new one. But I need that ramp.' This was at Port Isabel. We anchored [the ramp] onto the piers, and brought in a 'mule,' a little hoist, so when a boat came in, we would raise the ramp up; the boat would come under it, and we'd lower the ramp. We had to account for low and high tides, as to how much the ramp would drop.

"On the island side, there was nothing out there to land in—no pier. So we found a low spot at the south end of the island and took a barge and rammed it onto the land and loaded it with water to sink it so that it was fixed. We were still not in deep enough water, so we brought in a second barge and put it end to end to build the loading dock for the cars and to minimize the disturbance of the bay. The first barge had equipment on it—a loader and material. That loader dumped the material so you could walk on it. We put lights out there, run off a generator. So then we had both operations going." The entire process took two days.

The enormity of that accident underlined the launch of the new decade that would see controversy, upheaval, a push for new initiatives and new funding efforts, profound culture changes, and, in the words of one Texas senator, "friction and hostility" occur, many of these issues commencing under Mike Behrens's watch. For the seasoned engineer who had, like so many of his predecessors, graduated with his civil engineering degree from Texas A&M University, Behrens would need to prove himself a deft shepherd and caretaker indeed.

Michael Behrens addressing Short Course in 2006. (Michael Amador/TxDOT)

Mike Behrens. (Michael Amador/TxDOT)

A Contractor's Story

When Anne Weisman first met her husband-to-be, she saw far more potential in him than did his high school guidance counselor. "Johnny grew up doing construction work and started running machines at sixteen. He could earn more money on the road, as he says, than hoeing in a field of corn on the other side of the fence." The two were introduced by mutual friends during the summer before they turned sixteen; by then, Johnny was already hard at work. "In his family of five children," Anne remembers, "they got two pairs of blue jeans for Christmas, and socks, and their grandparents gave them underwear, and they got a couple of shirts. But they had to work for their other clothes, spending money, anything."

While Johnny was a senior in high school, his guidance counselor informed him that he was clearly not smart enough to enter college and should therefore seek a vocation that did not require a degree. "When we married," Anne says, "we wanted to get him off the road. So my dad got us jobs with the phone company. [Johnny] collected pay phones in the ship channel area of Houston, and I worked in the Yellow Pages department." Anne describes their early married life: "I'd been to college for two years but stopped to take a break [she was by then too deeply in love to focus on her studies] . . . and I wanted to go back to school. So Johnny got a job with the Highway Department. It was a huge concrete pour on a freeway somewhere

Johnny and Anne Weisman, principal owners of Hunter Industries. (Elizabeth Weisman/Hunter Industries)

Hunter Industries crew applies hot mix to SH 46 near Boerne. (Kevin Stillman/TxDOT)

in East Texas. He was a dispatcher for the concrete trucks. . . . He got a transfer and went to work for the Highway Department in Seguin, and I started back to school in the spring of '69. His mom was getting library certification in San Antonio, so he said to me, 'What would you think about me going to school?' I said, 'Most certainly.' So he took a couple of classes [while accompanying his mother on the drive from and to San Marcos]. He also took a couple of correspondence courses in drafting. . . . I worked at Sears part-time and was a waitress in a German restaurant also. He went to school two days a week, worked three days a week on the road for Dean Word Construction Company; on Friday nights, he unloaded boxcars at the lumberyard and worked there all day Saturday. It was fun. We'd ride our bicycles to his parents' place and help in the garden.

"When he graduated, he got a job at Leo P. Cloud Construction as the office manager. During that time, there was a woman there, Rosie, who was the brains (Rosie later came to work for us) who taught him real accounting —not 'book' accounting [such as artificial problems in textbooks], but the real [hands-on] accounting. . . . Rosie was wonderful. She and the other woman who worked there were divorced, raising children, and working for practically nothing (at that time the minimum wage was 85 cents an hour), so Johnny was always campaigning to get them both a raise."

Then an entrepreneur, the founder of Olmos Construction, invited Johnny and his brother to set up a construction business in Austin with his financial support. The proposed percentages were favorable to the Olmos owner, because he was a shrewd businessman. But he, too, saw a great deal of potential in Johnny Weisman. Meanwhile, both Anne and Johnny were determined to build a life and a business of their own through their steady efforts.

"Between me teaching school and Johnny working, on the weekends we would do site surveys for subdivisions on Nacogdoches Road in San Antonio, along with another guy from TxDOT," Anne said. "I would hold the rod, and they would shoot the grade and set pins for these streets. One day we came home and I looked down, and my jeans were moving: we had ticks all over us—our hair, everywhere. We pulled into the driveway, went straight into the laundry room, threw our clothes into hot water, and started picking seed ticks off each other. . . . [After the offer came to set up in business as the Austin arm of Olmos, they sold their new home for $5,000.] That was extra money that we could make on the side . . . , and that was exactly the number that this man needed from us. So we gave it to him. Johnny had already started going to Austin [Anne was still teaching school in New Braunfels]. Our first job was out on North Lamar, so on Saturdays I would come and dump the base trucks. You stand in the very back of the truck [as a guide], and they have their load of road base that they're dumping . . . and you back them up, and then you stop them, and they raise it, and then they lower their truck, and you give them the hand signals for whatever they do. Meanwhile the next guy is following behind you with the grader. I had this fabulous fake rabbit fur hat, because it was cold, it was winter, so I wore this fuzzy little hat, and these guys would then literally stop and look in their mirror, and look back, and go, 'That's a girl back there!' These days it could be anybody in a furry hat. But it was very surprising to them.

"My parents had told me that when I graduated from college, they would buy me a Trans-Am if I would stay home and go to the University of Houston. I said, 'No way.' So when we got out [of college], they said, 'We'll buy you a vehicle.' So I asked them to buy us a truck. It was a three-on-a-tree, standard, bare-bones white pickup truck, because we needed a company vehicle. And that truck stayed in our company—they would replace the universal joints and pass it down and recycle it. Sometimes Johnny would put a lien against the truck to make the payroll, because our guys had to get paid. Whether we got paid or not was a different story. We were not yet working with TxDOT. We did parking lots, apartment complex driveways— anything we could get to build up our backlog money, the TxDOT way. Mr. Zachry let us put our office on the Colorado River on 183, where the [Zachry Construction] yard still is today—Capitol Aggregates. We moved our job shack and porta potty there—that was a very exciting day. Then we bought a trailer—that was *really* amazing. I did the payroll."

From this modest bootstrap beginning, the Weismans gradually increased their operation, employee by employee, until before long they had one hundred workers on that same payroll. Soon they were working closely with TxDOT on a regular basis, constructing interstate segments, roads, and other projects within their areas of Texas. Eventually they bought out the Austin Olmos business, then went on to develop the companies they hold now: Hunter Industries, LTD, and Colorado Materials—which, combined, employ between five hundred and six hundred people.

The prosperity that the Weismans have generated extends to the families of these people they employ—people whose lives have been directly touched by the sense of purpose, hard labor, and dedication founded within the couple's own intimate partnership. In an industry that sees a far smaller number of women than men succeed, Anne has participated in tandem with her husband to contribute a strong and efficient cooperation with TxDOT. Johnny's policy of championing the welfare of his colleagues (such as Rosie the bookkeeper) has continued in his daily business practices; he knows each of his employees, maintains contact with them, demonstrates his personal concern for their welfare, makes a point of inviting them and their loved ones to his and Anne's ranch for hunting and recreation, and cooks their meals himself there. Together as a family, the Weismans built an empire from scratch—and in the truest sense, it is now an empire of families.

Without the vision of Dewitt Greer, who foresaw the wisdom of a system of "small-package" jobs made up of local "mom and pop" contractor companies with which to partner and build Texas roads, highways, and interstates rather than contracting huge corporate firms to do the jobs overall, the Weismans could not have created the successful companies they now own. They exemplify the best of Greer's idea, and the benefit to the state's economy, as well as those of local individuals, remains immeasurable.

The Hectic Decade: Mike Behrens, 2001–2007

Six months before the Queen Isabella Causeway disaster, Rick Perry, the man who had replaced George W. Bush in the Governor's Mansion, appointed his choices for the Texas Transportation Commission. Among them was Ric Williamson, the former legislator who was named commission chair nearly three years later. Throughout his six-year term as commissioner, Williamson provided a powerful thrust of agenda for transportation problems and solutions. According to one reporter, he also "had a knack for rubbing people the wrong way" and sometimes seemed to others as single aimed and ruthless as TxDOT's mightiest bulldozer.

The refusal of the legislature to raise the gas tax and/or address the funding deficit for transportation needs prompted other solutions. The prime example was the implementation of toll roads, one of which entailed a contract that broke all precedents and was let to a large, foreign-based company. That project, State Highway 130, generated controversy from its inception. Designed to relieve the staggering amount of truck traffic on I-35 between Texas and Mexico that had surged even more in the late 1990s, and intended to link a junction with I-35 in Georgetown to a route that extended all the way southward alongside Austin to Seguin, then southward and westward along I-410 to I-35 in southern San Antonio, the initial deal was sealed with Lone Star Infrastructure (a consortium made up of several major highway construction contractors and civil engineering firms) to design and build the first sections. Two further segments were awarded to the combined partnership of Cintra-Zachry. This contract gave that company the management of the road and the right to collect tolls for fifty years in a

Ric Williamson. (Kevin Stillman/ TxDOT)

The only 85-mile-per-hour speed limit in the United States is on this stretch of TX 130 between Austin and Seguin. (Kevin Stillman/TxDOT)

revenue-sharing agreement with the state. The state owned the road, and the company was responsible for financing, design, construction, operation, and maintenance over the life of the agreement. Although substantially a private-sector project, some costs for segments five and six, under the Cintra-Zachry contract, were borne by TxDOT. Cintra was a company based in Spain.

A second example of funds-seeking solutions was the scheme of breaking yet another longtime Highway Department precedent: borrowing the money for road construction through low-interest bonds. Some proponents argued that, on maturity, these bonds would demand less interest than the yearly inflation rate and therefore be cheaper in the long run than the old pay-as-you-go system in which prices rose through inflation before the project could be finished, thereby costing more than the original calculations. "Most old TxDOT hands agree that the pay-as-you-go policy is the best way to do things—if you can pay for it," said one of those same "old hands," adding, "but if the funding is not immediately available to complete important projects, the 'mega' projects that will generate more income, as well as the lesser, non-income-producing projects across the state, some feel that bonds are the solution."

Many Texas citizens strongly disagreed, seeing the burgeoning new debt and its servicing expenses as a burden that would weigh heavily on their children and grandchildren. Others countered this debate with the question, "Why shouldn't Texan children and grandchildren pay for the roads they'll be using? They will benefit from them; it's only right and proper that they should also assume the financial responsibility for them." Previous generations had paid their gasoline taxes to finance roads, and as the legislature had not yet raised that tax to accommodate new circumstances of inflation and lower gas mileage in modern vehicles, so, the reasoning followed, what the gas tax no longer covered could and should be supplemented through outside debt. Under Governor Rick Perry's fiscal policy, Texas began borrowing money in 2003 to pay for roads and owed $17.3 billion for them by the end of 2012, increasing the total outstanding state debt from $13.4 billion in 2001 to $37.8 billion in 2011, $43.53 billion for outstanding state debt in 2013, according to the Texas Bond Review Board (but $287 billion in actual total, making Texas' debt load the second largest in the nation after California's, as cited by a Reuters report), and $340.9 billion for total state debt as of 2014.

The third example was a proposal that would grow as infamous to its adversaries as it seemed attractive to certain political bodies: the Trans-Texas Corridor. The Trans-Texas Corridor was the revolutionary brainchild of Ric Williamson and symbolized the conflicts under which TxDOT struggled during the years between 2001 and 2010. Williamson and Governor Rick Perry proposed it in 2001. But it was Mike Behrens who would be expected to shepherd it, as well as the new toll roads, through the development and construction process.

Behrens expected the executive director position to be a demanding one from the beginning. On assuming it, he had stated, "We have a big system with a lot of miles, and of course there's a lot of needs out there. . . . That's no secret. There will be a lot of challenges in the years to come to meet the needs," both in terms of accommodating increasing traffic "and the big job of preserving the system we have." The process turned out to be even more of a roller-coaster ride than Behrens had anticipated: he was also up against the wishes and resentments of the citizens of Texas. Born and raised in Giddings, Behrens and his wife Karen had been glad to live for most of his career in Yoakum, where he served as, first, an engineering assistant for the district, then as an area engineer, a district planning engineer, an assistant district engineer, and eventually district engineer, and where they could raise their two small children in a similar small-town environment. Because of this, he was sensitive to the feelings of rural and small town residents.

Williamson did not live long enough to see the Trans-Texas Corridor shelved in 2009. In 2010, the official decision of "no action" was issued by the Federal Highway Administration, formally ending the project, and in 2011 the legislature canceled it through law. Its historical importance, however, is worthy of mention, both because of the departures from traditional policy that engrossed the energies, attention, time, and commitments of the Transportation Department throughout much of the decade and because the corridor perhaps presaged a future in which the department will have to grapple with these same issues again.

Proposed map for the Trans-Texas Corridor. (Courtesy of Texas State Library and Archives Commission)

Artist's concept of the completed corridor. (TxDOT Photo)

The plan involved a 4,000-mile network of super-corridors up to 1,200 feet wide to carry parallel links of tollways, rails, and utility lines from Oklahoma to Mexico and from east to west in South Texas. Its strategy was to redirect long-distance traffic to routes circumventing urban population centers and to provide stable corridors for future infrastructure improvements—such as new power lines from wind farms in West Texas—without the complicated legal procedures required to build on privately owned land. The tollway portion would have been divided into two separate elements: truck lanes and lanes for passenger vehicles, with usage fees charged to public and private concerns for utility, commodity, or data transmission within the corridor—in other words, basically creating another kind of toll road for utilities such as water, electricity, natural gas, petroleum, fiber-optic cables, and various telecommunications services. At the same time, the rail lines would serve several different functions: freight, commuter, and high-speed trains. Governor Perry proposed that it be partially financed, partially built, and wholly operated by private contractors who, in exchange for a multi-billion-dollar investment, would receive all toll proceeds, notably Cintra, the same company based in Spain that also contracted to build segments of SH 130, and its minority partner, the Zachry Construction Company, based in San Antonio. (Cintra, now defunct, was a spin-off of the much larger Ferrovial Group, created in 1998 to handle Ferrovial's global infrastructure business—chiefly toll roads and car parks—and currently reabsorbed by Ferrovial, which profits both from Cintra's lucrative toll collections and its declared bankruptcy, which removes its obligation to pay back its federally facilitated loans.)

The amount of land needed to achieve the specifications for the corridor was formidable: an estimated 584,000 acres. It was to be acquired through private property purchases and eminent domain right-of-way assertion by the state, for the management by private companies that would conduct day-to-day operations for some of the systems encompassed by the corridor. That the entire project was to be partly funded by private investors was an unheard-of step in an agency where pay-as-you-go state and federal funding had supplied and supported almost every previous construction at no additional charge to the public.

The Trans-Texas Corridor proved unpopular among many Texans, and protests occurred across the state. Here citizens marched to the Capitol in Austin for hearings. (Michael Amador/ TxDOT)

Private landowners, environmentalists, property rights activists, and people concerned about the volume of noise pollution that would potentially be generated within a mile of the corridor during peak hours: all these factions combined to produce such a large public outcry that, despite Perry's determined insistence for the project's completion, in 2009 TxDOT scrapped it. Instead, the department resorted to developing separate rights-of-way for road, rail, and other infrastructure using more traditional corridor widths for those modes. As the *Houston Chronicle* put it, the corridor schematics were "replaced with a plan to carry out road projects at an incremental, modest pace."

In January 2004, during Perry's fourteen-year period as governor, the number of transportation commissioners expanded from three to five, with Ric Williamson as chair. All five Perry appointees enthusiastically backed the Texas Trans-Corridor proposition. "Like it or not," Paul Burka wrote of Ric Williamson in *Texas Monthly* in June 2007, "the highway plan bears the Williamson trademark: It represents out-of-the-box thinking to solve a real problem, which is the inability of the revenue stream to keep up with the demand for mobility."

But for the first time in many decades, the general public of Texas no longer trusted its Transportation Department to deploy clean, nonpolitical practices with the public interest at heart. Faith had been shaken. Even tempered and equable as he was, Mike Behrens also served as the public face of TxDOT, the representative of all its works and plans. Big changes had been undertaken, starting with the initiatives of David Laney, the Transportation Commission chair, appointed in 1995 by Governor George W. Bush, who had created the first toll division within the Transportation Department. He supported efforts to allow the Texas Turnpike Authority, which owned the Dallas North Tollway, to be re-created as the North Texas Tollway Authority; he also had initiated the first toll projects in Central Texas, initiated TxDOT's shift toward innovative methods of highway finance (bonds), and implemented a $2.8 billion Rio Grande border infrastructure program. Laney had left the commission in 2001, making way for Williamson and the Trans-Texas Corridor vision he shared with his old friend and legislature colleague Rick Perry. And it was during this phase of big business privatization that, at least in the public's perceptions, the time-honored integrity of the Texas Highway Department began to disintegrate, bringing what the citizens of Texas saw as nearly a century's worth of apolitical striving, dogged honesty, hard work, and conscientious service full circle.

As Deputy Executive Director John Barton remembered, "We had the Sunset Review process by the legislature. . . . And so there was all this darkness and doubt around the agency, and it started to permeate our employee base. We all felt that we were trying hard, we were still the same agency we were before, we were still doing the same job, working hard for the public,

*Commission Chair Ric
Williamson and former
commissioner David
Laney. (Kevin Stillman/
TxDOT)*

but everybody hated us. We all wore our shirts with the TxDOT logo on it, and we didn't want to go to the grocery store because we didn't want people to know we worked for the agency. . . . In the general perception, we were maybe not corrupt, but politically manipulated maybe would be a better term"—but a term neither Gilchrist nor Greer would have ever tolerated in the old days. The Sunset Review process to which Barton referred was actually a double whammy: the Texas Sunset Advisory Commission ordered an evaluation of TxDOT the first time in 2008 because TxDOT was "an agency that had lost the trust of the public and the Legislature."

Williamson's sudden and untimely death on December 30, 2007, put an end to his forceful participation—although not a complete end to the controversy he had personally stirred and the communal disquiet he had fostered. Only fifty-five years old when he died of the third heart attack he had suffered since entering the Transportation Commission, Williamson was replaced by Bill Meadows of Fort Worth, who finished serving Williamson's original term (although not as chair, a position filled instead by new appointee and former Perry Chief of Staff Deirdre Delisi) and then departed, later becoming chair of the High Speed Rail Commission, newly organized in January 2014. In retrospect, the *Dallas Morning News* transportation columnist Rodger Jones wrote in his February 22, 2013, commentary that "Williamson . . . was the supremely confident architect of today's policy of leasing state right-of-way to private equity companies that would use their own money to build the roads, then collect tolls over decades and bank them as profits. He also helped hatch the ill-fated Trans-Texas Corridor project, ultimately labeled a greedy land grab."

A little earlier in 2007, in the midst of all this commotion and before Williamson's demise, Mike Behrens, having for six years helped the department navigate its way through the choppy ups and downs of its greatest controversies, retired from his thirty-five-year career with TxDOT, yielding the leadership to the commission's next choice.

Throughout department history, no matter what the political climate, the people of the Highway Department have often been first to respond to emergencies and natural disasters. Here a convoy of trucks from around the state heads into the zone devastated by Hurricane Ike in 2008. Similar, first-responder reactions occur during wildfires, tornadoes, flash floods, and traffic accidents. (Michael Amador/ TxDOT)

The first priority after a natural disaster such as Hurricane Ike is to secure victims and make sure they are safe. Then the department begins clearing roadways and putting the pieces back together. Many coastal employees worked long hours even while their own property received damage during the storm. (Michael Amador/ TxDOT)

Amadeo Saenz Jr., 2007–2011

Because of his capable handling of the Queen Isabella Causeway tragedy's aftermath, as well as his merit-worthy performance throughout his long career, Pharr District Engineer Amadeo Saenz was named the new executive director in 2007—the first person of Hispanic descent to hold the post. Steve Simmons, another long-term TxDOT career engineer who had served under Mike Behrens as the deputy executive director, agreed to stay on in that role when Saenz took the reins.

Amadeo Saenz Jr. grew up in Hebbronville, in Jim Hogg County, on his family's 1,400-acre ranch. His early proficiency at mathematics and fascination with engineering began during his ranch boyhood, where repairs and other kinds of problem solving required the talents and ingenuity he was developing. The TxDOT office in his hometown also awakened his curiosity: it would be a fine thing, he thought, to work there some day. After earning a bachelor of science degree in civil engineering with honors at the University of Texas at Austin in 1978, he immediately started his career for what would be, until his retirement thirty-three years later, his sole employer, as a TxDOT engineering laboratory assistant. Although he had more potentially lucrative job offers from within the private sector, TxDOT, Saenz felt, was the ideal fit for what he wished to accomplish: a chance to go beyond the theoretical processes he had studied in school (which would constitute most of the work in a corporation or large company) and instead implement those theories directly into tangible results.

This same enthusiasm had infused many engineers before him; the opportunities for actually building what they conceived and designed had always been a strong motivation among the department's professionals, and for sheer project diversity and mobility of workplace, no other employer could match what TxDOT delivered. Continuity, stability, and variety: these were the rewards that, for TxDOT engineers, superseded the chance for higher salaries, according to Saenz. Because his own interests extended to all fields, from the materials testing performed in the laboratories to drainage design to every level of road and bridge construction to ongoing maintenance after completion, he preferred to avoid the restrictions of specialization, instead learning as much as he could to master each area. In this way, too, he could inspect the work of others in situ in a district with an understanding of what each stage entailed: "In order to be able to review somebody else's work, you have to be able to design and build it yourself," Saenz explained. For this reason, he favored the more generalizing district placement rather than the specific focus of divisions.

Saenz became a licensed professional engineer in 1983, moving upward in the Pharr District office through the classic departmental progression to become an assistant area engineer, then assistant field operations engineer,

Almost twenty years before being named executive director, Amadeo Saenz received the 1988 Beautification Award from Lady Bird Johnson in Stonewall. (Geoff Appold/TxDOT)

area engineer, and deputy district engineer, before being named district engineer. Then, as the Queen Isabella Causeway repairs neared a close in November 2001, he received a new promotion: TxDOT's assistant executive director for engineering operations. His preference for well-rounded expertise and the extensive training and institutional knowledge he had received from his TxDOT mentors now paid off, to the state's benefit, with his appointment to the executive directorship.

By stepping into that post, he walked straight toward the center of the storm. The conflicts that had been kindled by toll roads and alternative funding methods now heated to greater intensity. Concern was expressed in the legislature about the choices made for the sake of transportation and about the direction in which TxDOT was heading. In a relatively short time after Saenz's appointment, the department was due for another Sunset Review. Meanwhile, Saenz went to work, establishing what he called a "One DOT" motto and policy, which he used to mend what he saw as a lack of cooperation between the agency's twenty-five districts. One of his most notable accomplishments was planning and initiating the renovation of I-35 from Waco to Salado—a prodigious undertaking that involved not only securing multiple of rights-of-way and preserving or relocating underground cables, utilities, and so forth but also the design and construction of overpasses and other added infrastructure, all to accommodate the widening and modernizing of what some have claimed is the entrance to the most highly trafficked segment of interstate in the country: the length from Georgetown through Austin and San Antonio, leading to Mexico.

Amadeo Saenz surveys the Texas coastal areas from a state plane prior to landfall of Hurricane Ike in 2008. (Kevin Stillman/TxDOT)

Then, on February 17, 2009, President Obama signed into law the American Recovery and Reinvestment Act of 2009 (Recovery Act). The main goals were "to create new jobs, maintain existing employment, spur economic activity, invest in long-term economic growth, and to foster unprecedented accountability and transparency in government spending." The passage of the Recovery Act brought a renewed focus on the importance of infrastructure throughout the United States. In addition to physical improvements to the national roadway system, the Recovery Act increased awareness of the need for preserving and improving highway infrastructure.

The State of Texas was given about $2.5 billion from the Recovery Act to put into highway and bridge projects, with a very short window of time in which to get them under construction, as the whole intention of the act was to quickly get people back to work after the recession. John Barton, as assistant executive director at that time, was given the job of managing that initiative for the department. They were able to accomplish this very successfully, Barton said, deploying projects that have since been "very impactful to the communities." For example, the Dallas–Fort Worth (DFW) connector project (which Barton called "critically important") was the largest Recovery Act–funded project in the nation. The billion-plus-dollar project utilized $260 million of Texas' Recovery Act funding to finish this project out. Nationally recognized as the most successful program in that particular initiative, it took everyone in the entire agency, plus others, as Barton explained, to carry it through cleanly and successfully. Construction on the project began in February 2010 to rebuild the corridor through Southlake, Grapevine, and the north edge of the DFW International Airport and reached substantial completion in November 2013. At its widest point on SH 114, the DFW connector now has up to twenty-four lanes, including fourteen main lanes, four toll lanes, and six frontage road lanes.

West 7th Street Bridge, Fort Worth

A popular east-west thoroughfare that crosses the Trinity River in Fort Worth connects downtown to growing mixed-used developments and the Cultural District. The project replaced the one hundred-year-old West 7th Street Bridge with a new signature structure—the world's first precast network arch bridge. The bridge designer was Dean Van Landuyt, who had come up with the concept previously and recognized that this was the right project for his innovative design.

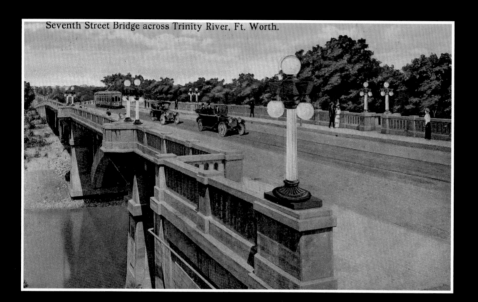

Seventh Street Bridge across Trinity River, Ft. Worth.

Postcard of the original West 7th Street Bridge in Fort Worth constructed in 1913. (Les Crocker Collection)

The new bridge consists of precast, post-tensioned concrete arches, floor beams, and deck panels and a cast-in-place concrete deck. The precast subcontractor for the floor beams was Heldenfels Enterprises, Inc., of San Marcos. The company manufactured 120, 82-foot beams in San Marcos and transported them to the bridge site. Since all of the elements were precast at separate locations, the demolition of the old structure and construction of the new bridge were completed in just five months. Here workers with Sundt Corporation pour concrete for one of twelve arches. (Michael Amador/TxDOT)

Once the arches had been cast, a delicate and technical operation began to move them into place. (Kevin Stillman/ TxDOT)

Tx DOT Bridge Designer Dean Van Landuyt. (Michael Amador/TxDOT)

Construction workers with Sundt Corporation building the deck. (Kevin Stillman/TxDOT)

When the $26 million bridge was finished in November 2013, it was a really big deal. Fort Worth Mayor Betsy Price and other state and city officials were on hand to cut the ribbon. (Kevin Stillman/TxDOT)

(Kevin Stillman/TxDOT)

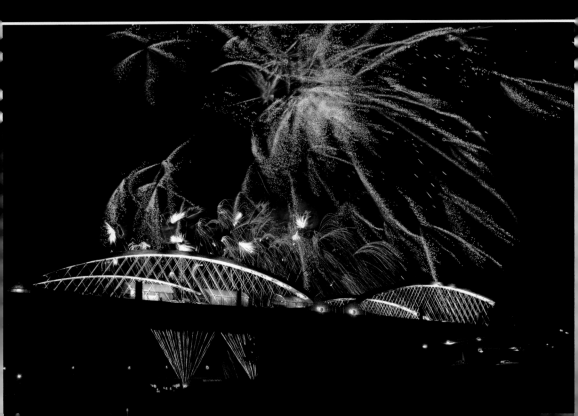

new bridge accommodates lanes of traffic with bicycle and estrian lanes on each side safely arated from traffic. The Fort Worth rict worked to make the planning execution a community project. rict Engineer Maribel Chavez and district staff received accolades awards for the department's rts. (TxDOT/Michael Amador)

An Engineer and His Road

*Chisholm Trail Parkway
entering Fort Worth.
(Courtesy of North Texas
Tollway Authority)*

When Burton Clifton attended the opening ceremony of the Chisholm Trail Parkway that now connects Fort Worth to Cleburne in May 2014, he felt a sense of humble satisfaction that rightfully could have equaled that of anyone else present. Looking out over the grand new toll road, he was gazing at the tangible culmination of fifty years of his life.

Clifton first entered the School of Engineering at Texas Tech University at the age of seventeen. At that time, like a number of his peers, he took a job working part-time for the Highway Department—in his case, in the Lubbock District office. But he was destined to complete only three semesters before discovering that the world had other plans for him. The year was 1944, and he had turned eighteen during the final month of 1943. Less than two months later, the US Army conscripted him into war service. "The Draft

Board didn't fool around back then," he said. "But we did have this advantage. By then they'd gotten to the point where they could send people where they belonged. I was an engineering student, so they sent me to the Corps of Engineers. There was a premed student with me at the same time—he went to the Medics. Carpenters and plumbers and electricians went to the Engineers. Everything was pretty rational [by then, so close to war's end]." Clifton spent the rest of his service building bridges, POW camps, hospitals, and latrines, in that order. After his release, he returned to Texas Tech on the GI Bill to complete his degree and resumed work part-time for the Lubbock District. His uncle was the assistant district engineer there, however, and protested Clifton's employment, so great was his fear of nepotism. "He didn't want to put a hex on me as being the boss's son," Clifton said. So after college, Clifton applied at other district offices to avoid discomfort for either of them.

By then he had a new young family, a wife and infant son. His first position was in Fort Worth; the second took him to Mineral Wells, surveying a road built earlier by the WPA. Many road-straightening projects followed in the hilly country around Palo Pinto, some as "windy as a snake" that needed reengineering and relocation to fit the land's contours, a job in which Clifton discovered great pleasure and enjoyment. Eventually, from acting as project engineer on these and on bridge constructions, he was promoted to resident engineer.

But the first full-scale project he ever worked on had been in Fort Worth, before his transfer to Mineral Wells. It would finally also be the last. The future corridor was then called the Southwest Freeway, and it was to be built at a time "when there wasn't anything else out there," said Clifton. "That was when I first got here to Fort Worth out of college. And I worked on that a lot—the location for it. Before we could get it built, though, we got preempted by a bunch of housing construction that moved part of the location. . . . Then after I came back from Mineral Wells, they already had the advanced planning done, and some other people were working on the detailed design of it." Yet more than fifty years would pass before that planning took final shape in asphalt and concrete.

Clifton went on to direct the Advanced Planning office of the Fort Worth District and helped plan most of the freeways, state highways, and interstates surrounding and running through Tarrant County and its environs. They were designed with extra-wide medians and extra-wide frontage, as Dewitt Greer dictated, with an eye always toward future expansion when traffic demands increased. Even so, according to Clifton, the foresight turned out not to be quite foresighted enough. For who could have prophesied the huge elevations in population and congestion to come?

Burton Clifton. (TxDOT Photo)

This was one reason that a venture fifty years in the planning and execution grew so valuable as a remedy. Others were funding, the growth of public involvement, environmental considerations and impact, bigger projects that usurped that station on the priority chain, and perceived need.

For comparisons, Burton Clifton could perhaps look backward to the Romans to find road development time frames similar to the one he helped nurse along. His employment with the department, which stretched to a total of fifty-five years—a record—was book-ended by the entrance and exit to the Chisholm Trail Parkway, a worthy memorial to his professional longevity and commitment. "Entire political careers have come and gone during that time," commented a reporter for the *Fort Worth Star-Telegram*. "And many of those folks thought they would never live to see the day when the road would open. Although it's a toll road, and not everybody is happy to have to pay to drive on it, it does hold the promise of a parkway—a place where you can feel good about traveling, a curvy road with hills, and lots and lots of trees. Beautiful scenery. Beautiful fencing." Five thousand trees were newly planted throughout its 28-mile length, and works of art were designed to mark the passage.

Burton Clifton attended the opening of the Chisholm Trail Parkway in 2014. Others on hand were district engineer former commissioner Meadows, TxDOT Executive Director Barton, former district engineer Chavez, and Victor Vandergriff Photo)

Another indication of why the project took so long, and how it reached completion at last, is its list of collaborators. The empty land that stretched along the route between Fort Worth and Cleburne remained largely vacant, despite the housing construction of which Clifton spoke. The desire for the road had to be strong enough to pull its supporters together. When it finally did, any disparities between their interests were transcended. The North Texas Tollway Authority; TxDOT; the North Central Texas Council of Governments (NCTCOG); Tarrant and John Counties; the cities of Fort Worth, Burleson, and Cleburne; and the Western Railroad and Union Pacific Railroad all combined forces toward this one goal: getting outliers into Fort Worth more easily and more pleasantly.

Customer quotes from the North Texas Tollway Authority website indicate they were pleased with the parkway: "A commute that normally takes 40 minutes took less than 20 minutes. . . . Once it is totally landscaped and complete it will be a perfect asset for Fort Worth, Tarrant County and Johnson County. A true work of art," said J. Jordan of Fort Worth. "I took the Chisholm Trail Parkway to work this morning, and it was fantastic. Thanks for building it for me! My normal commute takes me around 40 minutes. Today, I arrived at work about 10 minutes earlier. . . . Love it—got my TCU toll tag and everything," said Michael W. of Fort Worth.

And for Engineer Clifton Burton, who observed the opening ceremonies in pride of place, the job was worth it. Also were his lifelong efforts spent for TxDOT and Texas. "Everything I did was enjoyable," he said. "It was doing some good for the world."

(Michael Peters/TxDOT)

Sunset, Sunrise

Meanwhile, the controversies continued to roil, within as well as outside the department. In 2008, TxDOT went through the Sunset Review process; eleven of the mandatory twelve years had passed since the last one concluded with the legislature's approval. When the report came back with the recommendation that the five-person Transportation Commission be reduced to one person and contained (to quote Ben Wear, the transportation specialist for the *Austin American-Statesman*) "60 suggested course changes," the department was thrown into upheaval. Two years later, the same thing reoccurred. As Ben Wear wrote: "TxDOT was up for sunset review again because the 2009 Legislature failed to pass a sunset bill for the agency, which would have mandated certain changes and officially allowed the agency to continue operating for several years. Instead, the Legislature in a special session passed a law giving the agency another two years of life and mandating another sunset review." Although some of the flaws found by the first Sunset Review had since been addressed or were no longer relevant by the time the second one took place, the agency's preexisting structure was still required to undergo a revolution. (The 2009 legislature had created a new Department of Motor Vehicles, and thus ten of those suggestions, involving motor vehicle regulation, now concerned a different agency.) The second Sunset Review, accompanied by auditing firm Grant Thornton's negative analysis of the agency, prompted what the *Dallas Morning News* described as "a hand-picked panel of advisors" (in other words, the Restructuring Council formed by the commission) to make recommendations that would alter the course of TxDOT forever. One of these suggestions was to create a more "transparent, well-defined and understandable" system of deciding which projects to build with TxDOT money. Others included developing a TxDOT "code of ethics" (a throwback to J. C. Dingwall's self-auditing system) and creating an ethics hot line for reporting violations; strengthening prohibitions against lobbying at the state level by TxDOT employees (the agency would still be able to have federal lobbyists); improving the regulation of overweight and oversize vehicles; and requiring a license for outdoor advertising on rural roads.

But the most important and most far-reaching suggestion was for a drastic change in management.

The fact that Amadeo Saenz had risen to the position of executive director in the vortex of all the criticism directed toward the agency was an unfortunate piece of timing. According to the *Corpus Christi Caller-Times* of September 5, 2011, "Saenz announced his retirement in January after a scathing report from the agency's restructure council called for leadership changes. Deputy Executive Director Steve Simmons announced his retirement almost at the same time." The *Caller-Times* went on to say, "James

Bass, the department's chief financial officer, worked closely with Saenz in developing new models to finance transportation projects by partnering with the private sector. Although Saenz took over when the agency was being battered for promoting the Trans-Texas Corridor and for a $1 billion miscalculation in funding for construction, Bass said Saenz didn't let the public relations battles sway him. 'He knew we were implementing policy and at times a new direction and new policy might not always be the popular thing,' Bass said. 'But I think he had a strong belief it was for the benefit of the state and we were doing what truly was the right thing.'"

A pattern now seems established: throughout the years since the long reign of Dewitt Greer ended, a succession of executive directors has served for short bursts of three to six years each—one as long as seven years, and at least one for as brief a tenure as two. The heritage of purpose and dynamism originally fostered by Gibb Gilchrist and Dewitt Greer carried through ninety-two of these years in an unbroken line, linked as it was to the engineering profession and the engineers' habits of problem solving and practicality. After the Sunset Review process left the department with recommendations of shifts in management, stronger business-oriented

At a ceremony designating a portion of Interstate 69, Commissioner Jeff Austin, Deputy Executive Director John Barton, Commissioner Fred Underwood, and TxDOT Executive Director Phil Wilson. (Kevin Stillman/TxDOT)

approaches, and the mandate created by the legislature in 2009 that finally broke the precedent of the 1917 Texas Highway Act and removed its requirement that a professional engineer always head the department, a new breed of candidates appeared to fill the executive director's chair.

The first, Phil Wilson, was a former secretary of state for Governor Rick Perry. He earned a bachelor's degree from Hardin-Simmons University in Abilene and a master's degree in business administration from Southern Methodist University. Prior to his secretary of state role, he served as Governor Perry's deputy chief of staff and, before that, as his communications director for two years (he also led the Governor's Office of Economic Development and Tourism). Prior to that, he was an aid for nearly ten years to US Senator Phil Gramm.

According to a description of the Texas secretary of state's duties, in addition to handling election matters, the appointee acts as the governor's liaison with Mexico and the state's officer of protocol. The office serves as the formal repository for official and business records, publishes government rules and regulations, and attests to the governor's signature on official documents. In his role of secretary of state, Wilson served as Perry's designee on the Texas Enterprise Fund ($185 million) and the Emerging Technology Fund ($200 million). After leaving that position, he joined the electric energy company Luminant as vice president for public affairs before assuming his new employment as TxDOT chief, where he succeeded the last of the engineer chiefs, Amadeo Saenz Jr. Although he lacked experience in transportation, the administration of large bodies of employees engaged in disparate tasks (in this case, around twelve thousand), engineering, construction, and other related fields of endeavor, the five-person Perry-appointed Transportation Commission felt Wilson was a sound choice to push the agency forward toward the new business model urged by the Grant Thornton audit and report, which recommended the agency steer its leadership away from an engineering-dominated culture and hire more leaders with business management experience outside the department. His salary—$292,500—was the maximum amount allowed and the highest in the agency's history and was $100,000 per year more than that of his predecessor. (Interestingly enough, it exceeded Governor Perry's gross annual salary by more than $100,000—a proportionate comparison figure haling back to the earliest days of the department, when State Engineer George Duren earned one-fourth again more than his appointer, Governor James E. Ferguson, and the disparity between his salary and those of his Highway Department underlings was very pronounced indeed.) TxDOT employees found Wilson to be a ready listener and willing to reassess suggested changes that, ultimately, would merely confuse or complicate their jobs and their divisions further—especially those that might involve large-scale jigsaw readjustments and take many years to conclude.

As the business of infrastructure development has changed, innovative financing and idea partnerships have created revolutionary results. One such project is Klyde Warren Park, built over a depressed section of highway on the west side of downtown Dallas. The park was a private-public partnership. Pipeline magnate Kelcy Warren was one of the private contributors, and the park is named after his son Klyde. (Courtesy of Klyde Warren Park)

Thus, the change from the traditional engineer's approach to departmental purpose signaled an acute alteration in the agency's culture, structure, services, and objectives. During Wilson's two-year administration, several evolutions took place, particularly involving the outsourcing of services to private firms that had formerly been under the sole aegis of the state, via TxDOT. For instance, the IT Division was privatized, and its responsibilities were given to an overseas corporation. As described in the June 4, 2013, issue of the *Texas Tribune*, "The agency announced this week that it would transfer most of its IT functions to NTT DATA, a Japanese firm which has its North American headquarters in Plano. Most of TxDOT's current IT staff will be offered the chance to transfer to the company with the promise of at least six months of employment, according to TxDOT. About 350 TxDOT employees will be impacted. . . . The contract requires TxDOT to pay NTT DATA $190 million for five years of IT service, with the option to extend the contract for two more years."

In addition, the Maintenance Division equipment fleet was, if not dismantled, then significantly reduced in the Fleet Forward initiative. At least three thousand "assets" were auctioned off as either being too costly to repair or nearing an age when they would become costly to repair (the term "assets" includes both heavy road machinery and many vehicles). According to Darah Waldrip, TxDOT Fleet Forward information specialist, this modification to the old system also encompassed new software advances that had lately made a more centralized fleet division possible, more efficient, and cheaper to run, distributing equipment to districts across the state at need, along with mechanics. The fleet was therefore now separate from, but mutually reliant with, the Maintenance Division, and very large and heavy equipment that had not in the past been used as often would be rented from private suppliers when required.

Phil Wilson remained as TxDOT's executive director for a little over twenty-six months, from September 29, 2011, to December 18, 2013, at which time he resigned to become the new general manager and CEO of the Lower Colorado River Authority. The Texas Transportation Commission chose his replacement, Retired US Marine Corps Lieutenant General Joe Weber, after a deliberation that lasted until his hiring announcement on April 23, 2014.

The design and construction of the park was managed by Bjerke Management Solutions, and the design was led by two nationally recognized design firms, the Office of James Burnett and Jacobs Engineering Group, Inc. TxDOT selected Archer Western to construct the deck plaza. McCarthy Building Companies, Inc., served as the contractor to construct all of the amenities and complete the park. (Courtesy of Klyde Warren Park)

The park opened in 2012 and provides an urban oasis that connect a vital inner-city neighborhood with the Dallas Arts District and downtown. (Courtesy of Klyde Warren Park)

General Joe Weber receives a respectful welcome at the 2014 dedication of a new ferry operations building in Port Aransas. (Will van Overbeek/TxDOT)

Family within the Family

The San Angelo District of TxDOT was one of the six original districts established during the foundation of the Texas Highway Department. The second-largest district in the system, it encompasses more than 19,000 square miles, an area larger than the country of Switzerland. It includes the counties of Coke, Concho, Crockett, Edwards, Glasscock, Irion, Kimble, Menard, Reagan, Real, Runnels, Schleicher, Sterling, Sutton, and Tom Green. (Only the El Paso District, which exceeds the combined square mileage of Vermont and New Hampshire, supersedes it in size.)

Despite the fact that in the ten years from 2005 to 2015, the funding allocations for bid lettings on construction projects in the San Angelo District steadily climbed to triple its starting point, and that much of the maintenance budget was spent on repairing "Energy Sector Roads," which were crumbling (some down to "native soil") under the weight of rough oil industry traffic, the department leadership considered it expedient to cut back on personnel and facilities. All but two of the area offices were closed, with maintenance staff and equipment in the district's

Billy Chenault. (Karen Threlkeld/TxDOT)

eleven widely spaced maintenance barns trimmed down to a fraction of their former presence. This cutback followed the policy directives issued after the last Sunset Review repercussions.

Among the personnel in the San Angelo District, both current and retired, one theme can be heard over and over: the TxDOT family has proved to be far more real to them than a mere morale-building concept or form of lip service. In an isolated rural district that stretches over fifteen counties, the importance of departmental interconnection and goodwill has reigned paramount. This means that each person, from a part-time maintenance tech to the district engineer, always knew that he or she could pick up a telephone, call any other colleague from all the way across the state, and expect a spirit of reciprocal camaraderie, mutual understanding, and cooperation on whatever issue, problem, or friendly collaboration might arise. The sense that everyone employed by the department was "in it" together and that trust and a shared dedication to serve the public lay behind even the feistiest or joking workday rivalries,

Three generations: Dilan, Bobby, and Billy Chenault. (Karen Threlkeld/TxDOT)

made the cement that sealed department loyalties more firmly together than any other large agency's—possibly including those in the private sector. This applied to blood relations as well as a metaphorical family. Just as so many stories proliferate about fathers, daughters, sons, uncles, nephews, nieces, and cousins who have created a family tradition of graduating from engineering school and immediately taking employment in the engineering offices, so the same is true of the Maintenance Division.

One example of this trend of continuity is the Chenault family. With roots in Junction, Texas, that stretch back to pioneering times, for three generations the Chenaults have worked for TxDOT. William "Billy" Chenault joined the TxDOT force as a maintenance tech at the age of eighteen in 1956, on his second day out of the armed forces reserves. For thirty-seven years, he tended heavy equipment, mowers, pickup trucks, and office vehicles as a mechanic, while also using his array of multiple skills to tackle tasks from designing and welding elaborate labor-saving devices and life-saving safety railings to depositing hot mix, changing oil, and manning barrels of transmission fluid during his nine-hour workdays. In the coldest weeks of the Hill Country winter, he "cut brush until the eyelashes froze and then turned around and started sanding roads and bridges" to protect drivers from icy skids. His numerous commendations and certificates bespeak a lifetime of reliability. His son, Robert "Bobby" Chenault, followed in his father's work-boot prints at the age of seventeen and is now heading for forty years' steady service as a construction inspector. Bobby's son Dilan began as a maintenance tech with TxDOT at the age of twenty-six.

There are other examples of multigenerational families of employees through the years, but the Bohuslav family probably takes first prize. Led by former Yoakum district engineer Ben Bohuslav, there are at least nine others, including cousins, in-laws, and kids: Ken, John, Filecia, Thomas, James, Steve, Connie, Wanda and Sara . . . the Bohuslav highway family.

General Joe Weber.
(Michael Amador/
TxDOT Photo)

General Joe Weber, 2014–2015

Born in Weimar, Texas, Joe Weber grew up in Abilene and graduated from Texas A&M University in 1972, the same year as his classmate and friend, Governor Rick Perry. Later he earned a master's degree from the LBJ School of Public Affairs at the University of Texas. In his thirty-six years in the US Marine Corps, Weber served in numerous command and leadership positions throughout the United States and overseas, including tours in Europe, South America, Southeast Asia, and Iraq. He honed his management and leadership skills on and off the battlefield, where he trained, educated, and prepared thousands of marines, sailors, soldiers, and airmen for combat across the world theater. During those assignments, Weber also was responsible for the supervision of a variety of infrastructure projects that included road, port, aviation, and rail systems; he was also involved in their upkeep.

As the second non-engineer to hold the chief executive position at TxDOT, he brought a more traditional hierarchical approach back to the job, contrasting with his predecessor's business background, and also approved the hiring of two other former military officers as his chief strategy and innovations officer and chief of staff. Until his retirement in August 2015, Deputy Executive Director John Barton continued to bring his engineering expertise and longtime TxDOT career experience to the administration and oversight of all operations of the state's transportation system—in particular, those policies, programs, and operating strategies guiding cost-effective construction and maintenance; he also addressed the state's growing safety, mobility, and reliability needs.

On October 6, 2015, after eighteen months of service, Weber announced his retirement from the executive directorship of TxDOT. One of his final actions before leaving the post at the end of that calendar year was to sign a historic agreement, along with officers of the Federal Highway Administration and the Port of Corpus Christi, regarding the citizens of the Hillcrest and Washington-Coles neighborhoods (known collectively as the Northside Community in Corpus Christi), to mitigate TxDOT's decision to route access to the new Harbor Bridge through Northside, a design planned the previous spring. These two neighborhoods, originally segregated from the rest of the city by the restriction policies of the Jim Crow law, had remained home to their African American residents for many decades despite the increasing pressures and pollution of oil refineries hemming one side, the ship channel (walled by storage tanks and terminals) sealing off another, and Interstate 37 bordering a third. A civil suit brought on behalf of the residents was filed as a Title VI civil rights complaint; it pointed out that the two neighborhoods, long connected through mutual custom, would be severed from one another by the new "Red Route." In addition, Hillcrest would then become isolated from the rest of Corpus Christi—its schools, grocery stores, pharmacies, health amenities, and shopping conveniences—a familiar racial discrimination issue experienced throughout the years by many minority communities that had seen their homes and neighborhoods razed to make way for interstates and other transportation facilities. Upon investigation, the Federal Highway Administration agreed. Meanwhile, the Port of Corpus Christi wanted bridge construction to commence in a timely fashion.

TxDOT, which had thus far offered only a hike-and-bike trail and a new senior center as reparation for the damages to the dispossessed households, received the FHWA's stern insistence that the mitigation be much larger and include relocation and renters' funds as well as a number of other types of compensation. As a recipient of federal monies for the project, the department understood that a multi-million-dollar compliance reaching far beyond a usual eminent domain right-of-way settlement was imperative, and the four-party agreement, without precedent for any previous TxDOT project, was signed on December 17, 2015, two weeks before Weber's departure and the same day his successor, James Bass, was announced.

James Bass, 2016–

On January 1, 2016, James Bass stepped into the position of executive director of TxDOT, thus restoring the tradition of an insider chief with a long departmental history and a strong knowledge of the culture. Bass, a thirty-year agency veteran, first started working in the Fort Worth District during the summers while still in his teens and continued as a part-time engineering aide in the Austin District's South Travis/Hays County Area Office while earning an accounting degree at the University of Texas at Austin. He thus familiarized himself with some of the challenges and

*James Bass. (Michael Amador/
TxDOT Photo)*

problem-solving techniques involved in the engineering field as well as financial management.

After graduation, Bass took a job as an accounting clerk in the Finance Division's Revenue Accounting Section and before long ascended to the duties of budget analyst. In 1999, he assumed greater responsibility for the agency's budget as Finance Division director, and in 2005 he was appointed chief financial officer. During those years Bass was required, among other formidable tasks, to oversee the shift from the pay-as-you-go policy to the bond system, managing the department's newly acquired $20 billion debt and its yearly billion-dollar-plus debt service (an amount that constituted nearly 10 percent of TxDOT's annual budget). But by November 2015, the trend of financial burden turned when voters and legislators approved two constitutional amendments (one in the previous year) that designated another $4 billion of tax revenues to transportation—augmented by federal funds from a five-year transportation bill passed by Congress. Having navigated the currents of both approaches to TxDOT's massive funding demands, Bass was well informed of their complexities and in the best position possible to meld his long departmental loyalty and ethical discipline with practical strategies for the future.

The Approaching Centennial

In an interview, former deputy executive director John Barton summed up the past fifteen years' cultural transition, its ordeals, and its optimism, from an observational post within the department. The push to inaugurate the

Trans-Texas Corridor, he said was "not accepted by the community at large, was heavily covered by the media. A lot of negative perception of the agency [was] because of that. . . . In the last several years, that came to a closure. We responded to the Sunset Review and what the legislature wanted. And there had to be a change, and our former executive director Amadeo Saenz chose to retire, and that's when Mr. Wilson came on. And so there was an opportunity to have a course change. We said, 'Okay. We have this bad stuff was behind us, the Trans-Texas Corridor is not happening, Chairman Williamson is no longer our chairman. We have a new chairman, we have a new executive director, and let's see if we can go in a different way.'"

Now, in the final decade before the Centennial year of the old Texas Highway Department's creation, this "different way" has included fewer internal design/build intentions (relying instead on outside design/build consultants and contractors), more private engineering firms and financiers, more outside-hire project inspectors, more external maintenance dependence, and more tollways. It has embraced other branches of outsourcing that in turn have occasionally promoted more controversy and even scandal, such as overpriced and overcharged contracts with external consultants that lie beyond the traditional TxDOT oversight by the guards protecting its carefully audited integrity. A good example of this was a certain information system contractor who billed TxDOT a total of $1.65 million over a five-year period, or an average of more than $300,000 a year—compared to TxDOT's Executive Director General Joe Weber, who earned $273,000 a year, not including his benefits. Another problem that surfaced in 2015 was the mishandling of toll collections, particularly for the tollways surrounding Austin, when Xerox, the corporate contractor managing the accounting, inherited mistakes from the previous contractor that caused more than thirty thousand people who had already purchased toll tags covering their mileages to get double-billed—with ongoing, escalating penalties. The differences in priorities between state agency supervision and private business blaze through these discrepancies.

It has also witnessed a change in practices for the workforce itself. Throughout the two new-millennial decades, the duties of district engineers have shifted dramatically for major as well as minor ventures. "What I've seen [starting in approximately 2000] is an evolution in Dallas–Fort Worth and the big metro areas across the state," said Fort Worth District Public Information Officer Jodi Hodges. "The DEs have started playing a major role in the negotiations for these billion-dollar projects. It's beyond engineering; it's financials; it's legal agreements. I don't know how many thousands and thousands of hours they've pored over agreements and had conference calls with Austin to hammer out the DFW Connector and the North Tarrant Express, and the LBJ and I-35. It's so much more than just designing these roads."

FM 170, the River Road, tracks near the Rio Grande and through the mountainous desert country between Study Butte and Presidio west of Big Bend National Park. It is one of the most beautiful drives in Texas. (J. Griffis Smith/TxDOT)

In November 2014, the Texas voting public endorsed a statewide bill, Proposition One, to divert half of the general revenue from the state's oil and gas taxes destined for the Economic Stabilization Fund, also known as the Rainy Day Fund, and to redesignate it to provide for the repair and maintenance of roads—with the express caveat that it not be used to build toll roads. For several years prior, the condition of a number of rural roads, especially those in the fracking-centered oil and gas production areas of the Eagle Ford Shale in South Texas and the Barnett Shale in North Central Texas, had degraded to such a degree due to heavy drilling equipment and oil industry traffic that emergency measures were taken to keep the infrastructure even moderately usable, such as stripping off the remaining asphalt and converting (or reverting) thoroughfares to gravel surfaces. These steps were taken because there was not sufficient money with which to renew the roads to their best-level paving. In addition, the number of traffic deaths had risen exponentially from one year to the next in the regions affected by the fracking activities, in part exacerbated by the large increase in the numbers of cumbersome vehicles, in part by the hazardous conditions themselves. Nearly 80 percent of voting Texans approved the Proposition One funding measure, a clear indication that they recognize the importance of maintaining current roads and constructing new ones: the work TxDOT—the Texas Highway Department—was originally created to do and will continue to do, as long as it has the monies and the specialized and dedicated staff with which to execute its mission.

The Present

Just as roads have always changed human ways of life, economies, and our contact with the worlds that lie waiting beyond our own village borders, so they also inflect our minds and spirits. Roads help civilizations prosper; this fact remains indisputable. But the other ways in which they change us are subtler, muting our sense of vast distances, collapsing our former perceptions of time and space via the rate at which we traverse them. We readily admit that some day, many years ahead, roots, vines, soil, and stone will probably engulf our buildings. What we may not notice is how a good, "fast" road erases that same natural world from our eyes now. Our ever-increasing speed eclipses the clouds overhead, the meadows and deserts and mountains and forests, into a glimpsed blur that instantly dissolves, ending any chance for close and thoughtful knowledge. And from the sky, intimacy is impossible: all we receive is the big picture, a rolling map. Most of us, however, gladly embrace this trade-off for efficiency and convenience.

Smudge pot. (TxDOT Photo)

The future holds many alternatives for human and goods conveyance that will transform the way we move from one place to another and perhaps one day render highways obsolete. Some of these are currently on the drawing board. Some are in physical development; a few are already actualized: self-driving cars, freight shuttles, aerial mass-transit gondolas, hyper-loops, both real and imagined; high-speed rail hurling us across the prairies; virtual travel experienced from our armchairs. But the purpose of this narrative is to show us where we have been rather than to speculate upon where we may be going. History is the diary of a journey, of how we have arrived at where we stand now, at this milestone moment. It carries within it the meaning of the past and the shape of the future—mutable, adjusting to the topography as its pavers go forward.

Acknowledgments

We thank the following people for their invaluable assistance, expertise, and support in the development of our tribute to the Texas Highway Department (known today as TxDOT). We could not have created this book without you.

Anne Cook, librarian of the TxDOT Photo Library, who provided more information, guidance, source material, ideas, suggestions, and patience than any librarian should ever be expected to in a lifetime.

Liz Clare, whose knowledge of the Texas Highway Department's early history is equaled only by her proficiency in visual archeology and her indefatigable research.

Hong Yu, Texas Transportation Institute librarian, who has so patiently and generously fielded our requests.

Jill Crane, cataloging librarian for the Louis J. Blume Library at St. Mary's University, who provided map information on the Old Spanish Trail.

Brian Barth
John Barton
James Bass
Mike Behrens
Sarah Bird
Thomas Bohuslav
Jon Brumley
Rebecca Brumley
William G. Burnett
Tracy Cain
Mary Cearley
Mirabel Chavez
Robert Chenault
William Chenault
Betsy Christian

George Christian
Dennis Christiansen
Margaret Crisp
Ed Davis
Jon Engelke
Patti Everitt
Gregg Freeby
Janelle Gbur
Henry Gilchrist
Lucille Goode
Marc Goode
Jodi Hodges
John Huddleston
Tom Johnson
John Kelly

Debbie Koehler
Robert and Elyse Lanier
Bobby Littlefield
Greg Malatek
Tom Mangrem
Tim McClure
Walter McCullough
Matt McGregor
William Meadows
John Miller Morris
Lana Nelson
Willie Nelson
Arnold Oliver
Lawrence Olsen
Karen Othon
Michael Peters

Doug Pitcock
Richard Ridings
Geoff Rips
Walt Robertson
Amadeo Saenz
Steven E. Simmons
Karen Threlkeld
Gary Trietsch
Charles Walker
Joseph Weber
Anne Weisman
Johnny Weisman
Martha Strain Wilkinson
Jolynne Williams
Carol O'Keefe Wilson
Marcus Yancey

Sponsor Credits

Gold Sponsor
Williams Brothers Construction Company, Inc.

Centennial Sponsor
Zachry Construction

Diamond Sponsor
AECOM
Dannenbaum Engineering
Holt Cat
Hunter Industries, Ltd
J. D. Abrams
J. H. Strain & Sons
Kelcy Warren

Transportation Friend
AGC of Texas
Heldenfels Enterprises
JACOBS
Bob & Elyse Lanier
Michael Baker International
Texas Transportation Institute

Century Club

Michael W. Behrens, P.E.

David Eric Bernsen

Mark J. Bloschock, P.E.

Rick Collins, P.E.

Rita & Delvin Dennis, P.E.

Billie Jean & Bobby Evans, P.E.

GSD&M

Ned Holmes

Ted Houghton Jr.

Johnny Johnson

David M. Laney

Pati & Bill Meadows

Michael & Dianna Noble, P.E.

Hans C. "Chris" Olavson, P.E.

Toni & Arnold Oliver, P.E.

Kirby Pickett, P.E.

Mary Lou Ralls, P.E.

Becky & Luis Ramirez, P.E.

Amadeo Saenz Jr., P.E.

Steven E. Simmons, P.E.

Camille, Vicki, and Henry Thomason, P.E.

Gary K. Trietsch, P.E.

Roger G. Welsch, P.E.

HNTB

In the foreword, by Willie Nelson:

Texas

Words and Music by Willie Nelson

(c) 1989 FULL NELSON MUSIC, INC.

All Rights Controlled and Administered by EMI LONGITUDE MUSIC

All Rights Reserved International Copyright Secured Used by Permission

Reprinted by Permission of Hal Leonard Corporation

To the excellent staff of Texas A&M University Press, you have our eternal gratitude: our peerless editor, Shannon Davies; her sharp-eyed aide Emily Seyl; Patricia Clabaugh, associate editor and our tireless shepherd and project manager; Gayla Christensen, marketer extraordinaire; Kevin Grossman, design and production assistant manager, who makes it all happen on the page; and Cynthia Lindlof, the fabulous copyeditor who has waded through our manuscript to produce a pristine text.

To all the men and women of TxDOT: we salute you for the essential and dedicated service, so often taken for granted, with which you improve and protect the lives of Texans and other human beings everywhere.

And finally, to our wonderful partners and spouses, Anne Cash Edwards, Jeffery Poehlmann, and Elizabeth Cogwin: thank you for seeing us through.

When I retired from TxDOT in January 2012, I embarked on telling this story. In my career I wrote about, worked with, and observed the thousands of dedicated men and women employees of TxDOT to provide a world-class transportation system for a very large and complicated state. I am indebted to the zeal and enthusiasm from Carol Dawson, who took on this project without previous transportation experience. She quickly understood and passionately embraced just how powerful and dynamic this story was and has brought it to life in this book. And the insights of my longtime TxDOT collaborator and friend Geoff Appold, who pored through thousands of images to find those in this book to tell the story before us, also shaped the arc of this story. Without the close partnership with Carol and Geoff, this story could not have been told. And finally, I am eternally thankful for the support and encouragement of my wife, Cash.

—Roger Polson

I will be forever grateful to Roger Polson for suggesting this project to me, providing such excellent insight and guidance about what research paths to take and which rocks to look under, and in general, exposing me to a whole new world that I, like so many members of the public, had previously taken for granted. I also thank Geoff Appold for his inveterate eye and know-how in ferreting out images that could show everything I was trying to tell.

—Carol Dawson

I was immediately on board when Roger came up with this ambitious undertaking. We knew one thing for sure: In the TxDOT archives, containing well over a half million photographic images and documents, we would discover more than we could ever hope for to illustrate a hundred-year history, and we did; reducing it to the number included in this book was indeed a challenge. We had two things in our favor: my having worked closely with the TxDOT photographic staff, some of the most talented photographers in this state, for nearly 30 years; and super-librarian Anne Cook, who came to work for the department in 1989 and worked under my supervision to organize the already massive collection and to ensure its preservation. Contents of this collection include not only a running history of TxDOT but also Texas travel images used in travel publications, especially in *Texas Highways* magazine.

—Geoff Appold

Appendix

Highway/Transportation Commission Members, 1917–2015

Meeting date	Name	Position	Hometown	Service dates
6/4/17	Curtis Hancock	Chairman	Dallas	6/4/1917–3/20/1919
	Thomas R. McLean		Mt. Pleasant	6/4/1917–1/21/1918
	H. C. Odle		Meridian	6/4/1917–1/21/1918
1/24/18	Curtis Hancock	Chairman		
	R. M. Hubbard		New Boston	1/22/1918–3/20/1919
	J. G. Fowler		San Antonio	1/22/1918–11/22/1918
11/25/18	Curtis Hancock	Chairman		
	R. M. Hubbard			
	C. S. Fowler		San Antonio	11/23/1918–2/21/1921
3/20/19	R. M. Hubbard	Chairman		3/21/1919–2/5/1925
	C. S. Fowler			
	C. N. Avery		Austin	3/21/1919–2/21/1921
2/23/21	R. M. Hubbard	Chairman		
	D. K. Martin		San Antonio	2/22/1921–2/15/1925
	W. W. McCrory		San Antonio	2/22/1921–6/20/1923
7/16/23	R. M. Hubbard	Chairman		
	D. K. Martin			
	G. D. Armistead		San Antonio	6/21/1923–9/10/1924
11/24/24	R. M. Hubbard	Chairman		
	D. K. Martin			
	John B. Bickett Sr.		San Antonio	11/22/1924–10/8/1926
2/26/25	Frank V. Lanham	Chairman	Dallas	2/16/1925–12/3/1925
	Joe Burkett		Eastland	2/16/1925–12/3/1925
	John B. Bickett Sr.			
12/14/25	Hal Moseley	Chairman	Dallas	12/4/1925–10/8/1926
	John M. Cage		Stephenville	12/4/1925–10/8/1926
	John B. Bickett Sr.			

Highway/Transportation Commission Members, 1917–2015

10/19/26	Eugene T. Smith	Chairman	San Antonio	10/9/1926–1/31/1927
	Scott Woodward		Fort Worth	10/9/1926–1/31/1927
	George P. Robertson		Meridian	10/9/1926–1/31/1927
2/14/27	R. S. Sterling	Chairman	Houston	2/1/1927–10/6/1930
	Cone Johnson		Tyler	2/1/1927–3/17/1933
	W. R. Ely		Abilene	2/1/1927–10/6/1930
10/20/30	W. R. Ely	Chairman		10/7/1930–6/5/1933
	Cone Johnson			
	D. K. Martin			10/7/1930–2/13/1937
6/19/33	John Wood	Chairman	Timpson	6/6/1933–2/14/1935
	W. R. Ely			6/6/1933–2/14/1935
	D. K. Martin			
2/1/35	Harry Hines	Chairman	Wichita Falls	2/15/1935–2/13/1937
	John Wood			2/15/1935–5/18/1939
	D. K. Martin			
2/25/37	Robert Lee Bobbitt	Chairman	San Antonio	2/14/1937–5/18/1939
	Harry Hines			2/14/1937–4/11/1941
	John Wood			
6/22/39	Brady Gentry	Chairman	Tyler	5/19/1939–3/25/1945
	Harry Hines			
	Robert Lee Bobbitt			5/19/1939–2/14/1943
4/22/41	Brady Gentry	Chairman		
	Robert Lee Bobbitt			
	Reuben Williams		Dallas	4/12/1941–2/19/1947
12/16/43	Brady Gentry	Chairman		
	Reuben Williams			
	Fred E. Knetsch		Seguin	2/15/1943–3/6/1949
3/26/45	John S. Redditt	Chairman	Lufkin	3/26/1945–12/31/1948
	Reuben Williams			
	Fred E. Knetsch			
3/18/47	John S. Redditt	Chairman		
	Fred E. Knetsch			
	Fred A. Wemple		Midland	2/20/1947–1/5/1949
1/27/49	Fred A. Wemple	Chairman		1/6/1949–2/25/1951
	Fred E. Knetsch			
	A. F. Mitchell		Corsicana	1/5/1949–2/25/1951
3/28/49	Fred A. Wemple	Chairman		
	A. F. Mitchell			
	Robert J. Potts		Harlingen	3/7/1949–3/15/1955
2/26/51	E. H. Thornton Jr.	Chairman	Galveston/ Houston	2/26/1951–5/20/1957
	Fred A. Wemple			2/26/1951–2/15/1953
	Robert J. Potts			

Highway/Transportation Commission Members, 1917–2015

2/23/53	E. H. Thornton Jr.	Chairman		
	Robert J. Potts			
	Marshall Formby		Plainview	2/16/1953–5/20/1957
4/25/55	E. H. Thornton Jr.	Chairman		
	Marshall Formby			
	Herbert C. Petry Jr.		Carrizo Springs	3/16/1955–3/15/1959
5/30/57	Marshall Formby	Chairman		5/21/1957–3/15/1959
	Herbert C. Petry Jr.			
	Charles F. Hawn		Athens	5/21/1957–4/16/1963
3/25/59	Herbert C. Petry Jr.	Chairman		3/16/1959–7/17/1967
	Charles F. Hawn			
	Hal Woodward		Coleman	3/16/1959–7/17/1967
4/17/63	Herbert C. Petry Jr.	Chairman		
	Hal Woodward			
	J. H. Kultgen		Waco	4/17/1963–6/27/1968
7/31/67	Hal Woodward	Chairman		7/18/1967–6/27/1968
	H. C. Petry Jr.			7/18/1967–6/10/1973
	J. H. Kultgen			
7/10/68	J. H. Kultgen	Chairman		6/28/1968–3/31/1969
	H. C. Petry Jr.			
	Garrett Morris			7/10/1968–5/18/1971
4/1/69	Dewitt C. Greer	Chairman	Austin	4/1/1969–11/16/1972
	Garrett Morris		Fort Worth	
	H. C. Petry Jr.			
5/31/71	Dewitt C. Greer	Chairman		
	H. C. Petry Jr.			
	Charles E. Simons			5/19/1971–11/16/1972
12/11/72	Charles E. Simons	Chairman	Dallas	11/17/1972–6/10/1973
	D. C. Greer			11/17/1972–3/23/1981
	H. C. Petry Jr.			
6/27/73	Reagan Houston	Chairman	San Antonio	6/11/1973–4/30/1979
	D. C. Greer			
	Charles E. Simons			6/11/1973–11/21/1978
11/30/78	Reagan Houston	Chairman		
	D. C. Greer			
	A. Sam Waldrop		Abilene	11/22/1978–5/20/1979
5/21/79	A. Sam Waldrop	Chairman		5/21/1979–7/15/1981
	D. C. Greer			
	Ray A. Barnhart		Pasadena	5/2/1979–1/31/1981
3/19/81	A. Sam Waldrop	Chairman		
	D. C. Greer			
	Robert H. Dedman		Dallas	3/19/1981–7/15/1981

Highway/Transportation Commission Members, 1917–2015

4/23/81	A. Sam Waldrop	Chairman		
	Robert H. Dedman			
	John R. Butler Jr.		Houston	3/23/1981–7/24/1985
7/16/81	Robert. H. Dedman	Chairman	Dallas	7/16/1981–5/17/1983
	A. Sam Waldrop			7/16/1981–3/30/1983
	John R. Butler Jr.			
3/30/83	Robert. H. Dedman	Chairman		
	Robert C. Lanier		Houston	3/30/1983–5/17/1983
	John R. Butler Jr.			
5/19/83	Robert C. Lanier	Chairman		5/18/1983–2/24/1987
	Robert. H. Dedman			5/18/1983–7/24/1985
	John R. Butler Jr.			
7/25/85	Robert C. Lanier	Chairman		
	Thomas M. Dunning		Dallas	7/25/1985–4/28/1986
	Ray C. Stoker Jr.		Odessa	7/25/1985–3/13/1991
4/29/86	Robert C. Lanier	Chairman		
	Robert M. Bass		Fort Worth	4/29/1986–2/24/1987
	Ray C. Stoker Jr.			
2/25/87	Robert H. Dedman	Chairman		2/25/1987–3/13/1991
	Robert C. Lanier			2/25/1987–7/28/1987
	Ray C. Stoker Jr.			
7/29/87	Robert H. Dedman	Chairman		
	John R. Butler Jr.			7/29/1987–4/23/1989
	Ray C. Stoker Jr.			
4/26/89	Robert H. Dedman	Chairman		
	Ray C. Stoker Jr.			
	Wayne B. Duddlesten		Houston	4/24/1989–11/19/1991
4/26/91	Ray C. Stoker Jr.	Chairman		3/14/1991–11/19/1991
	Robert H. Dedman			3/14/1991–11/19/1991
	Wayne B. Duddlesten			
11/21/91	Ray Stoker Jr.	COT*		11/20/1991–1/10/1993
	Henry R. Munoz III		San Antonio	11/15/1991–11/25/1994
	David Bernsen		Beaumont	11/15/1991–1/10/1993
1/26/93	David Bernsen	COT		1/11/1993–4/19/1995
	Henry R. Munoz III			
	Anne S. Wynne		Austin	1/11/1993–3/11/1999
3/31/94	David Bernsen	COT		
	Anne S. Wynne			
	Ruben R. Cardenas		McAllen	3/4/1994–1/3/1995
4/27/95	David M. Laney**	COT	Dallas	4/20/1995–4/27/2000
	David Bernsen			4/20/1995–5/20/1997
	Anne S. Wynne			
5/29/97	David M. Laney	COT		
	Anne S. Wynne			

Highway/Transportation Commission Members, 1917–2015

	Robert L. Nichols		Jacksonville	5/21/1997–6/30/2005
3/26/99	David M. Laney	COT		
	Robert L. Nichols			
	John W. Johnson		Houston	3/12/1999–4/27/2000
4/28/00	John W. Johnson	COT		
	David M. Laney			4/28/2000–4/11/2001
	Robert L. Nichols			
4/26/01	John W. Johnson	COT		4/28/2000–1/28/2004
	Robert L. Nichols			
	Ric Williamson		Weatherford	4/21/2001–1/28/2004
1/29/04	Ric Williamson	COT		1/29/2004–12/30/2007
	John W. Johnson			1/29/2004–1/8/2007
	Robert L. Nichols			
	Hope Andrade		San Antonio	12/15/2003–1/28/2008
	Ted Houghton Jr.		El Paso	12/15/2003–10/6/2011
6/30/05	Ric Williamson	COT		
	Hope Andrade			
	Ted Houghton Jr.			
	John W. Johnson			
1/25/06	Ric Williamson	COT		
	Hope Andrade			
	Ted Houghton Jr.			
	Ned Holmes		Houston	1/8/2007–5/31/2012
	Fred Underwood		Lubbock	1/8/2007–2/12/15
1/31/08	Hope Andrade	COT		1/29/2008–4/30/2008
	Ted Houghton Jr.			
	Ned Holmes			
	Fred Underwood			
5/29/08	Deirdre Delisi	COT	Austin	4/30/2008–10/6/2011
	Ted Houghton Jr.			
	Ned Holmes			
	Fred Underwood			
	Bill Meadows		Fort Worth	4/30/2008–2/1/2013
10/27/11	Ted Houghton Jr.	COT		10/7/2011–2/12/2015
	Ned Holmes			
	Fred Underwood			
	Bill Meadows			
	Jeff Austin III		Tyler	10/20/2011–
6/28/12	Ted Houghton Jr.	COT		
	Fred Underwood			
	Bill Meadows			
	Jeff Austin III			
	Jeff Moseley		Houston	

Highway/Transportation Commission Members, 1917–2015

4/25/13	Ted Houghton Jr.	COT		
	Fred Underwood			
	Jeff Austin III			
	Jeff Moseley			
	Victor Vandergriff		Arlington	3/26/2015–
2/13/15	Tyron D. Lewis	COT	Odessa	2/13/2015–
	Jeff Austin III			
	Jeff Moseley			
	Victor Vandergriff			
	J. Bruce Bugg Jr.		San Antonio	2/13/2015–

* Chairman of transportation (COT); title changed 11/20/91

** Appointed to commission 4/3/1995–4/19/1995 and made chairman 4/20/1995

Selected Bibliography

The *Dallas Morning News* articles are arranged earliest to most recent.

Dallas Morning News

"Available Funds for State Highway Commission's Work in 1918 Expected to Approximate Two Billion Dollars," Oct. 14, 1917.

Duren, George. "State Highway Engineer in Favor of Paved Roadway from Denison to Galveston, via Dallas and Houston," Oct. 14, 1917.

"Confer Here Feb. 5 on State Highways; Roads Serving Army Camps Will Be Given the Preference," Jan. 31, 1918.

Duren, George. "Says Training Schools for Women Public Necessity," July 7, 1918.

"State Highway Representatives Get Concessions at Washington," Oct. 13, 1918.

"Road Work Contracts to Be Given on Bid System," May 11, 1919.

"Corsicana Needs Hotel Facilities: Ambition Is to Become Headquarters for Supplies and Oil Men," Oct. 1, 1921.

"Urge Bonnet Song for 'MA' Ferguson," Aug. 12, 1924.

"May Name S. B. Moore as Consulting Engineer," Mar. 29, 1925.

"Governor to Open New Nueces Bridge," Apr. 19, 1925.

"Single Day's Pardon Record Made by Mrs. Ferguson When She Signs 36 Proclamations," July 15, 1925.

"Claims Barred from Highway Body Minutes," Sept. 19, 1925.

"Road Lettings Are Defended," Oct. 18, 1925.

"Book Contracts Not Approved by Dan Moody," Oct. 21, 1925.

"Moody Again before Jury," Oct. 21, 1925.

"Hank Denies Sanction," Oct. 23, 1925.

"Governor Gets into the Controversy," Oct. 26, 1925.

Thornton, William M. "Satterwhite and Fergusons at Parting of Ways after Fight over Special Session," Oct. 27, 1925.

———. "'Ferguson Mad with Power'—News of Split Confirmed as Road Inquiry Progresses," Oct. 28, 1925.

"Ask Resignations Tendered," Nov. 23, 1925.

"Mr. Moody's Triumph," Nov. 23, 1925.

Thornton, William M. "Demand Governor Call Legislature: Insist upon Highway Investigation despite the Resignation of Two," Nov. 24, 1925.

Acheson, Sam. "$286,025 Road Check to State," Dec. 2, 1925.

"Award Is Made for Rebuilding Part of Pike," Dec. 15, 1925.

"Officials to Open Invisible Track Road," Dec. 15, 1925.

"Mrs. Ferguson Opens Road: New Type Highway Inspected by Governor and Her Party," Dec. 16, 1925.

Thornton, W. M. "Highway Head to Quit Place on January 15," Dec. 31, 1925.

"Bond Ruling Is Far-Reaching," Jan. 6, 1926.

Thornton, William M. "Road Bonds Approval to Be Held Up," Jan. 6, 1926.

Goodwin, Mark L. "Moody to Ask Rehearing as to Road Law," Jan. 21, 1926.

Vinson, Curtis. "Hoffman Firm Witnesses for Defense Heard," Jan. 31, 1926.

"Convicts in Texas Prison on Increase," Feb. 22, 1926.

"A. C. Love New State Engineer," Mar. 6, 1926.

"Plans Made for Opening Bridge over Red River," Apr. 12, 1926.

"Road Bonds Bill Framed," May 6, 1926.

"Moody Attacks Road Location," June 11, 1926.

"Is Mr. Ferguson for Repudiation?," June 17, 1926.

"Ferguson and Moody Differ on Validation," June 22, 1926.

"Yett-Nalle Case Closed," June 26, 1926.

"Committee Questions Highway Board Officials and Obtains Mass of Figures on Details," Sept. 24, 1926.

"Ferguson Demanded 10 PerCent on Bid, Says Contractor," Oct. 19, 1926.

"Contractors Testify They Paid for Ads in Ferguson Forum," Oct. 20, 1926.

"Nephew of Jim Ferguson Offered to 'Splite Commission,' Witness Informs Investigators," Oct. 21, 1926.

"Tells Probers of Road Deal," Nov. 12, 1926.

"Probers Inspect Bell County Road," Dec. 15, 1926.

"Moody Opens New Bridge over Trinity," Sept. 3, 1927.

"Moody Helps to Open New River Bridge: Louisiana Joins Hands with Texas at Sabine Memorial," Nov. 12, 1927.

"Moody Smiles, Convinced He Will Win Out," July 27, 1928.

"Board Stands by Gilchrist," Sept. 26, 1928.

"Moody's Road Plan Favored," Oct. 22, 1928.

"Patrol Forms Available," Aug. 27, 1929.

"Two States Plan Three New Bridges," Oct. 2, 1929.

"Moody Asks Support for Bridge Measures," Dec. 10, 1929.

"Austin-San Antonio Road to Be Improved," Apr. 2, 1930.

"Federal Highway Work in Texas Will Proceed," Aug. 22, 1930.

"Gaps in Highways to Be Completed Next Year," Dec. 20, 1930.

"State Winner in Bridge Suit for $165,000," Jan. 16, 1932.

"Frank Denison Is Turned Down by Senate Vote," Feb. 9, 1933.

"Work in Senate Is Retarded by Denison Battle," Feb. 12, 1933.

"Denison May Demand Office Right Monday," Feb. 27, 1933.

"Governor's Appeal for Denison Vote Refused by Senate," Mar. 1, 1933.

Thornton, William M. "Denison Wants Job Only If All Sides Satisfied," Mar. 2, 1933.

"Denison's Road Warrant Order Upset by Allred's Decision That Ely Is de Facto Board Chairman," Mar. 4, 1933.

"Vote of Senators Is Ruled Immune in Denison Action," Mar. 14, 1933.

"Commission Seeks Money for Roads, Not Reforestation," Mar. 29, 1933.

"Denison in Court Again Asking Seat on Highway Board," May 6, 1933.

"Denison Denied Mandamus Plea," May 18, 1933.

"Frank L. Denison Again Is Refused Road Board Seat," May 30, 1933.

"Highway Endorsement," May 30, 1936.

Fee, J. E. "Sales Tax Best to Pay Pensions, Sanderford Says," July 14, 1936.

Wasson, Alonzo. "State Faces Dire Crisis in Pension Fund," Sept. 29, 1936.

"Jim Ferguson's Brother Jailed for Contempt in $300,000 Squabble," May 19, 1940.

"Engineer Position Unaffected by Kinship," Aug. 4, 1940.

"Famed Texas Builder of Bridges Dies," Nov. 28, 1943.

"Bryan Y. Cummings, Noted Lawyer, Dies," Nov. 25, 1950.

"Death Takes Mrs. Duren," Aug. 17, 1958.

Interview. Texas Governor Rick Perry. Transcript, Jan. 23, 2009.

Lindenberger, Michael A. "Dallas Lawyer David Laney Inducted into Texas Transportation Hall of Honor," Oct. 6, 2010.

———. "TxDOT Chief Executive Amadeo Saenz to Resign," Jan. 26, 2011.

Jones, Rodger. "Ghost of Ric Williamson Could Haunt Hunt for Texas Highway Money," Feb. 22, 2013.

Other Periodicals

"Amadeo Saenz, Son of South Texas, Leaves State's Top Transportation Post." *Corpus Christi Caller-Times*, Sept. 5, 2011.

Batheja, Aman. "Bass Picked as New Head of Transportation Agency." *Texas Tribune*, Dec. 15, 2015.

———. "Bullet Train Failed Once, but It's Back." *New York Times*, Mar. 6, 2014.

———. "Details Murky on State's Largest Contracts." *Texas Tribune*, Apr. 4, 2015.

———. "Erroneous Toll Bills Fuel Criticism at Hearing." *Texas Tribune*, Feb. 11, 2015.

———. "For TxDOT, a $2 Billion 'Perception Problem.'" *Texas Tribune*, June 11, 2012.

———. "House, Senate Leaders Confirm Transportation Funding Deal." *Texas Tribune*, May 26, 2015.

———. "In Legislature, Toll Roads Facing Strong Opposition." *Texas Tribune*, Mar. 22, 2015.

———. "Musk: Texas a 'Leading Candidate' for Hyperloop Track." *Texas Tribune*, Jan. 15, 2015.

———. "Perry Lobbying for A&M Official to Take Over TxDOT." *Texas Tribune*, Mar. 31, 2014.

———. "Phil Wilson: The TT Interview." *Texas Tribune*, Aug. 23, 2012.

———. "Texas Funding Woes Nab Spotlight." *Texas Tribune*, May 20, 2014.

———. "TxDOT Outsourcing IT Operations to Private Firm." *Texas Tribune*, June 4, 2013.

"Bill Meadows to Take Helm of New Texas High-Speed Rail Commission." *Fort Worth Star-Telegram*, Jan. 24, 2014.

Blumenthal, Ralph. "Protection Sought for Vast and Ancient Incan Road." *New York Times*, June 18, 2014.

Brown, David, and Rhonda Fanning. "How Drug Smugglers Are Taking Advantage of the Texas Oil Boom." *Texas Standard*, May 22, 2014.

Burka, Paul. "The Farm to Market Road." *Texas Monthly*, Apr. 1983.

———. "Ric Williamson Suffers Fatal Heart Attack." *Texas Monthly*, Dec. 30, 2007.

———. "The Tea Party, Rick Perry, and Debt." *Texas Monthly*, Apr. 18, 2013.

California Freight Mobility Plan. "Trend Analysis: Farm-to-Market." *Cal-Trans*, 2013. http://dot.ca.gov/hq/tpp/offices/ogm/CFMP/Dec2014/Appendices/Appendices/Appendix_I_Freight_Trends/Appendix_I-2_TrendAnalysis_FarmToMarket_121114%20.pdf.

Carter, O. K. "The Biggest Reason for Arlington's Decades-Long Boom." *Fort Worth Star-Telegram*, Aug. 2007.

Cassidy, Jon. "Tell <apos>Em I Ain't Got It: Texas Debt Hits $341 Billion." Watchdog.org Texas Bureau, Jan. 17, 2014. http://watchdog.org/124193/tell-em-aint-got-texas-debt-hits-341-billion/.

Christian, Carol. "Bragging Rights or Embarrassment? Katy Freeway at Beltway 8 Is World's Widest." *Houston Chronicle*, May 13, 2015.

Cortes, Maria. "Joe Battle." *El Paso Times*, Oct. 9, 1991.

Cox, Mike. "Getting My TV Kicks on 'Route 66.'" *Texas Co-op Power Magazine*, May 2012.

———. "The Haunting of the Old Travis County Jail." *Texas Escapes Online Magazine*, Oct. 14, 2010. http://www.texasescapes.com/MikeCoxTexas Tales/Haunting-of-Old-Travis-County-Jail.htm.

———. "Queen Isabella Causeway Disaster." *Transportation News*, Sept. 2002.

Crow, Kirsten. "Harbor Bridge Buyout Agreement Finalized." *Corpus Christi Caller-Times*, Dec. 17, 2015.

Dannin, Ellen. "Crumbling Infrastructure, Crumbling Democracy: Infrastructure Privatization Contracts and Their Effects on State and Local Governance." *Northwestern Journal of Law and Social Policy* 6, no. 1 (Winter 2011): 47–105.

Davies, Alex, and Vivian Giang. "The World's 11 Wildest Highways: The Katy Freeway Is the Widest in the World." *Business Insider Magazine*, Aug. 29, 2012.

Duren, George A. "The Cost of a Mile of Road." *Highway Engineer and Contractor*, May 1919.

Ewing, Cortez A. M. "Proceedings of Investigation Committee, House of Representatives Thirty-Fifth Legislature: Charges against Governor James E. Ferguson Together with Findings of Committee and Action of House with Prefatory Statement and Index to Proceedings." *Political Science Quarterly* 48, no. 2 (June 1933): 184–210.

"First Automobile Trip in Texas." Forney Historic Preservation League. *Dallas Morning News*, Oct. 6, 1899. http://www.historicforney.org/archives/articles/firstautotrip.html.

Fulton, Candace Cooksey. "Heald Named A&M Outstanding College of Engineering Alumnus." *Brownwood Bulletin*, May 5, 2002.

Good Roads Magazine, vol. 8, 1907; vol. 52, 1917.

"Government Salary Explorer." *Texas Tribune*, 2014. http://salaries.texastribune.org/.

Gubbels, Jac. "Texas Landscapes for Safety: The Psychologic Approach to Landscape Planting." *Landscape Architecture* 59, no. 2 (1940) 63–65.

Hannaford, Alex. "Highway Injustice: Texas Leads the Nation in Unsolved Highway Serial Homicides." *Texas Observer*, Nov. 7, 2012.

Harris, Dilue Rose. "The Runaway Scrape: The Non-combatants in the Texas Revolution; Reminiscences." *Quarterly of the Texas Historical Association* 4 (1901): 155–89.

"Highway Commission Gets First Plates." *Fort Worth Star-Telegram*, Aug. 5, 1917.

"The History of El Camino Real de los Tejas." TexasCounties.net, Oct. 29, 2015. http://www.texascounties.net/articles/el-camino-real-de-los-tejas/history.htm.

Jackson, Tom. "Behind the Installation of the World's First Precast Network Arch Bridge in Fort Worth." *Equipment World*, May 29, 2013.

Landuyt, Dean Van. "West 7th Street Bridge: Creating a New Gateway in Fort Worth." *Aspire Magazine*, Fall 2013.

Lewis, Arthur H. "Hetty Green's Railroad." *Railroad Magazine*, Oct. 1963.

Long, Trish. "Trish Long: A Look Back at El Paso's Trans-Mountain Road." *El Paso Times*, Nov. 28, 2011.

Malanga, Steve. "Deep in the Debt of Texas: Local Liabilities Threaten the State's Fiscal Reputation." *City Journal* (Spring 2013). http://www.city-journal.org/2013/23_2_snd-texas-debt.html.

McGraw, Al. "*Origins of the Camino Real in Texas.*" *Texas Almanac, 2002–2003.* http://texasalmanac.com/topics/history/origins-camino-real-texas.

McSwain, Ross. "Crazy Jim, Goodeye Helped Establish New Time on Stage Run from San Angelo to Ozona." *San Angelo Standard-Times*, Aug. 23, 1964.

Mosqueda, Priscila. "A Neighborhood Apart." *Texas Observer*, June 1, 2015.

Neuharth, Al. "Traveling Interstates Is Our Sixth Freedom." *USA Today*, June 22, 2006.

Nicholson, Eric. "Sorry, Fort Worth, but Your New 'Signature Bridge' Is Pathetic." *Dallas Observer*, Oct. 9, 2013.

Nijhuis, Michelle. "What Roads Have Wrought." *New Yorker*, Mar. 20, 2015.

Nolen, Oran W. "By Stage from Corpus Christi to San Antonio." *The Cattleman*, Jan. 1946.

Owen, Sue. "Texas near Top in Local Debt per Capita." *Austin American-Statesman*, Dec. 6, 2013.

Ramshaw, Emily. "Perry Helps Advisers Take a Big Next Step." *New York Times*, Dec. 8, 2012.

Ravo, Nick. "Francis C. Turner, 90, Dies; Shaped the Interstate System." *New York Times*, Oct. 6, 1999.

"Rediscovering and Preserving El Camino Real. TexasCounties.net, Oct. 29, 2015.http://www.texascounties.net/articles/el-camino-real-de-los-tejas/preservation.htm.

Satija, Neena. "In 1917, Similarities to Gov. Rick Perry's Indictment." *Texas Tribune*, Aug. 17, 2012.

Smith, Griffin, Jr. "The Highway Establishment and How It Grew and Grew and Grew." *Texas Monthly*, Apr. 1974.

State of Texas. "Penal Code: Title 8. Offenses against Public Administration: Chapter 36. Bribery and Corrupt Influence." *Journal of the Senate of the State of Texas.* Regular Session of the Seventy-Second Legislature Convened Jan. 8, 1991; Adjourned May 27, 1991.

Sturtevant, Rebecca. "Historic Agreement Resolves Environmental Justice Complaint in Corpus Christi, Texas." Lawyers' Committee for Civil Rights under Law, press release, Dec. 18, 2015.

Sweaney, Brian D. "The Next Four Years." *Texas Monthly*, Dec. 2014.

Taibbi, Matt. "Rick Perry: The Best Little Whore in Texas." *Rolling Stone Magazine*, Oct. 26, 2011.

"10 States with Enormous Debt Problems: Report." Huffington Post; Reuters, Aug. 28, 2012.

Texas Highway Department. "Fifty: Golden Anniversary Edition, Texas Highway Department 1917–1967." *Texas Highways Magazine*, 1967.

"TxDOT Director Behrens Announces Plans to Retire from Post at End of August." *Lubbock Avalanche-Journal*, June 1, 2007.

Tyler, R. G. "Earth Roads: Roads and Pavements." *University of Texas Bulletin* no. 1922 (Apr. 15, 1919).

"Want Duren Retained." *Fort Worth Star-Telegram*, May 10, 1919.

Wear, Ben. "Sunset Easy on TxDOT in Latest Review." *Austin American-Statesman*, Nov. 19, 2010.

———. "Top Officials at TxDOT See Sharp Jump in Pay." *Austin American-Statesman*, July 7, 2012.

———. "TxDOT Finance Chief James Bass Named to Lead Agency." *Austin American-Statesman*, Dec. 17, 2015.

Weingroff, Richard F. "Firing Thomas H. McDonald—Twice." *Highway History*, Federal Highway Administration, updated Nov. 18, 2015. https://www.fhwa.dot.gov/infrastructure/firing.cfm.

———. "The Genie in the Bottle: The Interstate System and Urban Problems, 1939–1957." *Public Roads Magazine* 64, no. 2 (Sept./Oct. 2000). https://www.fhwa.dot.gov/publications/publicroads/00septoct/urban.cfm.

———. "The Greatest Decade: 1956–1966." *Highway History*, Federal Highway Administration, updated Nov. 18, 2015. https://www.fhwa.dot.gov/infrastructure/50interstate.cfm.

———. "Three States Claim First Interstate Highway." *Public Roads Magazine* 60, no. 1 (Summer 1996). https://www.fhwa.dot.gov/publications/publicroads/96summer/p96su18.cfm.

Wheeler, Camille. "Roadside Memories." *Texas Co-op Power Magazine*, July 2007.

"Who's Who on Highway Board." *Fort Worth Star-Telegram*, June 20, 1917.

Publications

American Oil and Gas Historical Society. "Kokernot Oil Company." 2015. http://aoghs.org/stocks/kokernot-oil-company/.

Austerman, Wayne R. *Sharps Rifles and Spanish Mules: The San Antonio–El Paso Mail, 1851–1881*. College Station: Texas A&M University Press, 1985.

Austin, Stephen F. *Papers: Notes and Sketches*. Austin, Texas State Archives.

Bedell, Barbara Fortin. *Colonel Edward Howland Robinson Green and the World He Created at Round Hill*. South Dartmouth, MA: Self-published, 2003.

Biggers, Don H. *Our Sacred Monkeys, or 20 Years of Jim and Other Jams*. Austin: Self-published, 1933.

Blodgett, Dorothy, Terrell Blodgett, and David L. Scott. *The Land, the Law, and the Lord: The Life of Pat Neff*. Austin: Home Place Publishers, 2007.

Borth, Christy. *Mankind on the Move: The Story of Highways*. Washington, DC: Automotive Safety Foundation, 1969.

Brown, Norman D. *Hood, Bonnet, and Little Brown Jug: Texas Politics 1921–1928*. College Station: Texas A&M University Press, 1984.

Caro, Robert. *The Years of Lyndon Johnson: The Path to Power*. New York: Alfred A. Knopf, 1982.

Carr, William G. *School Finance*. Stanford, CA: Stanford University Press, 1933.

Carter, Kathryn Turner. *Stagecoach Inns of Texas*. Austin: Eakin Press, 1994.

Clare, Liz. *From Pioneer Paths to Superhighways: The Texas Highway Department Blazes Texas Trails 1917–1968*. Texas State Library and Archives Commission, updated Nov. 14, 2011. https://www.tsl.texas.gov/exhibits/highways/kingshighways/page1.html.

Clark, Ira G. *Then Came the Railroads: The Century from Steam to Diesel in the Southwest*. Norman: University of Oklahoma Press, 1958.

Coghlan, Byron Kemp. "The Organization of a State Highway Department for the State of Texas." Master's thesis, University of Illinois, 1916.

Conan, Michael, ed. *Environmentalism in Landscape Architecture*. Vol. 22. Washington, DC: Dumbarton Oaks Research Library and Collection, 2000.

Cruse, Stephen Douglas. "Roads for Texas: Creation of a State Highway Department." Master's thesis, University of North Texas, 1992.

Davidson, Janet F., and Michael S. Sweeney. *On the Move: Transportation and the American Story*. Washington, DC: Smithsonian Institution and National Geographic, 2003.

Eisenhower, Dwight D. *Report on Trans-continental Trip*. Rock Island Arsenal, Nov. 3, 1919. Eisenhower Archives, Dwight D. Eisenhower Presidential Library, Abilene, Kansas.

———. "Through Darkest America with Truck and Tank." In *At Ease: Stories I Tell to Friends*, 155–168. New York: Doubleday, 1967.

Ely, Glen. "Riding the Western Frontier: Antebellum Encounters on the Butterfield Overland Mail, 1858–1861." Master's thesis, Texas Christian University, 2005.

Erlichman, Howard J. *Camino del Norte: How a Series of Watering Holes, Fords, and Dirt Trails Evolved into Interstate 35 in Texas*. College Station: Texas A&M University Press, 2006.

Faulkner, William. *The Reivers*. New York: Random House, 1962.

Federal Bureau of Investigation. "Highway Serial Killings: New Initiative on an Emerging Trend." Apr. 6, 2009. https://www.fbi.gov/news/stories/2009/april/highwayserial_040609.

Fehrenbach, T. R. *Lone Star: A History of Texas and the Texans*. 1968. Reprint, Cambridge, MA: Da Capo Press, 2000.

Gilchrist, Gibb. "Gibb Gilchrist: An Autobiography." Unpublished, Dec. 23, 1987.

Gillette, Michael L. *Lady Bird Johnson: An Oral History*. Oxford: Oxford University Press, 2013.

The Gospel of Good Roads: A Letter to the American Farmer. New York: League of American Wheelmen, 1891.

Gredler, Sara, Rick Mitchell, and Megan Ruiz. *A Historic Context for Texas Roadside Parks and Rest Areas, Texas Roadside Parks Study*. Austin: Mead & Hunt, for Environmental Affairs Division Historical Studies Branch, TxDOT, 2013.

Gubbels, Jac. *American Highways and Roadsides*. Boston: Houghton Mifflin, 1938.

Hagan, Hilton. *An Informal History of the Texas Department of Transportation*. Austin: Texas Department of Transportation Public Information Office, 1991.

"History." Terrell Chamber of Commerce Convention & Visitors Bureau. Accessed Dec.11, 2015. http://www.terrelltexas.com/history.

Hofsommer, Donovan L. *The Southern Pacific, 1901–1985*. College Station: Texas A&M University Press, 1986.

Huddleston, John David. "Good Roads for Texas: A History of the Texas Highway Department, 1917–1947." PhD diss., Texas A&M University, 1981.

Jackson, Mary. "Greer Building." Austin: Tourist Information Center, Texas State Capitol Building, n.d.

Lewis, Arthur H. *The Day They Shook the Plum Tree*. New York: Harcourt Brace, 1963.

Lewis, Tom. *Divided Highways: Building the Interstate Highways, Transforming American Life*. Ithaca, NY: Cornell University Press, 2013.

"Main Andean Road: Qhapaq Ñan." UNESCO World Heritage Centre, 2015. http://whc.unesco.org/en/qhapaqnan/.

Mann, Charles C. *1491: New Revelations of the Americas before Columbus*. New York: Knopf, 2005.

Masterson, V. V. *The Katy Railroad and the Last Frontier*. Norman: University of Oklahoma Press, 1952.

McClure, Tim, and Roy Spence. *Don't Mess with Texas: The Story behind the Legend*. Austin: Idea City Press, 2006.

McMurtry, Larry. *Roads: Driving America's Great Highways*. New York: Simon and Schuster, 2000.

Moody, Ralph. *Stagecoach West*. New York: Thomas Y. Crowell, 1967.

Morehead, Richard. *Dewitt C. Greer, King of the Highway Builders*. Austin: Eakin Press, 1984.

Muñoz, Ascencion. *Tuberculosis Turned El Paso into a Health Center*. Vol. 19. El Paso: El Paso Community College Libraries, 2000.

National Park Service. "National Register of Historic Places." Updated Nov. 2, 2013. http://focus.nps.gov/nrhp/SearchResults/.

Ormsby, Waterman L. *The Butterfield Overland Mail: Only through Passenger on the First Westbound State.* Edited by Lyle H. Wright and Josephine M. Bynum. Berkeley: University of California Press, 1991.

Overton, Richard C. *Burlington Route: A History of the Burlington Lines.* New York: Knopf, 1965.

———. *Gulf to Rockies: The Heritage of the Fort Worth and Denver-Colorado and Southern Railways.* 1953. Reprint, Westport, CT: Greenwood, 1970.

Patterson, B. D., G. Ceballos, W. Sechrest, M. F. Tognelli, T. Brooks, L. Luna, P. Ortega, I. Salazar, and B. E. Young. 2007. Digital Distribution Maps of the Mammals of the Western Hemisphere, version 3.0. NatureServe, Arlington, Virginia, USA.

Potts, Charles S. *Railroad Transportation in Texas.* Austin: University of Texas, 1909.

Reed, S. G. *A History of the Texas Railroads.* 1941. Reprint, New York: Arno, 1981.

Richards, Anne. "Anne Richards's Inaugural Address." Regular session of the Seventy-Second Legislature convened January 8, 1991.

Scannell, Jack C. *A Survey of the Stagecoach Mail in the Trans-Pecos 1850–1861.* Lubbock: West Texas Historical Association Yearbook, 1971.

Schmitt, Angie. "The Revolving Door: TxDOT's Phil Wilson, 'Revolver in Chief.'" *Streetsblog U.S.A.,* Feb. 1, 2013. http://usa.streetsblog.org/2013/02/01/the-revolving-door-txdots-phil-wilson-revolver-in-chief/.

Sitton, Thad. *The Texas Sheriff: Lord of the County Line.* Norman: University of Oklahoma Press, 2000.

Slack, Charles. *Hetty: The Genius and Madness of America's First Female Tycoon.* New York: Ecco , 2004.

Slotboom, Erik. *Houston Freeways: A Historical and Visual Journey.* Cincinnati: Oscar F. Slotboom, Publisher, 2003.

Slotboom, Oscar. *Dallas-Fort Worth Freeways: Texas-Sized Ambition.* Cincinnati: Oscar Slotboom, Publisher, 2014.

Smith, Marian. "The Stagecoach in Travis County." Manuscript, n.d., Austin History Center, Austin, Texas.

State of Texas General Laws. "House Bill No. 2, State Highway Department." Approved Apr. 4, 1917.

Stilgoe, John. "Roads, Highways, and Ecosystems." Harvard University, National Humanities Center, July 2001. http://nationalhumanitiescenter.org/tserve/nattrans/ntuseland/essays/roads.htm.

Strand, Ginger. *Killer on the Road: Violence and the American Interstate.* Austin: University of Texas Press, 2012.

The Texas Constitution. Vol. 10, *The Laws of Texas 1822–1897.* Austin: Gammel, 1898.

Texas Department of Transportation. "Chief Financial Officer: James Bass." Accessed Dec. 27, 2015. http://www.txdot.gov/inside-txdot/adminis tration/chief-financial.html.

———. *Gulf Intercoastal Waterway: Legislative Report 82nd Legislature.* Austin: TxDOT, n.d.

———. *The History of Texas License Plates: 80th Anniversary Edition.* Austin: TxDOT, 1999.

———. *The History of the Texas State Highway Department.* Austin: Texas Department of Transportation, 1938.

———. Internal memo to employees, Dec. 15, 2015.

———. *Transportation Funding: Understanding Transportation Funding in Texas.* Austin: TxDOT, 2014–15.

Thonhoff, Robert H. *San Antonio Stage Lines, 1847–1881.* El Paso: Texas Western Press, 1971.

Tocqueville, Alexis de. "Why the Americans Are So Restless in the Midst of Their Prosperity." In *Democracy in America,* vol. 2. London: Saunders and Otley, 1840.

Updegrove, Mark K. *Indomitable Will: LBJ in the Presidency.* New York: Crown Publishers, 2012.

Villa R., Alfonso. "The Causeways of Yucatan." In *Yaxuna-Cobá Causeway. Contributions to American Archaeology* 9 (Apr. 1934): 189–92.

Wallace, Karl Edward, III. "Texas and the Good Roads Movement: 1895 to 1948." Master's thesis, University of Texas at Arlington, 2008.

Warren, Robert Penn. *All the King's Men.* New York: Harcourt, Brace, 1946.

Weingroff, Richard F. "The League of American Wheelmen." In *The Road to Civil Rights,* 9–13. Washington, DC: Federal Highway Administration, 2012.

Wilson, Carol O'Keefe. *In the Governor's Shadow.* Denton: University of North Texas Press, 2014.

Zlatkovich, Charles P. *Texas Railroads.* Austin: University of Texas Bureau of Business Research, 1981.

Interview Subjects and Consultants

Brian Barth (Fort Worth district engineer)

John Barton (former deputy executive director)

Michael Behrens (former executive director)

William Burnett (former executive director)

Tracey Cain (district engineer, San Angelo District)

Maribel Chavez (former district engineer, Abilene, Fort Worth)

Robert Chenault (construction inspector, Junction, San Angelo District)

William Chenault (retired maintenance tech, Junction, San Angelo District)

Burton Clifton (former resident engineer, Mineral Wells, and advanced planning director, Fort Worth)

Anne Cook (TxDOT photo librarian)

Gregg Freeby (head of Bridge Division)

Henry Gilchrist (son of Gibb Gilchrist)

Marquis Goode (former executive director)

Jodi Hodges (Fort Worth public information officer)

Joe Holley (journalist)

Tom Johnson (executive vice president, Texas Association of General Contractors)

John Kelly (former district engineer, San Antonio)

Debbie Koehler (secretary to Tom Johnson, AGC)

Robert Lanier (former highway commissioner)

Janet Lea (ad campaigns; Sherry Matthews Advocacy Marketing)

Bobby Littlefield (Waco district engineer)

Tom Mangram (former area engineer, Alpine)

Sherry Matthews (ad campaigns; Sherry Matthews Advocacy Marketing)

Tim McClure (Don't Mess with Texas—GSD&M)

Walter McCullough (former district engineer, San Angelo District)

William Meadows (former highway commissioner—current high-speed rail commissioner)

Larry Norwood (journalist)

Arnold Oliver (former executive director)

Lawrence Olsen (executive vice president, Texas Good Roads and Transportation Association)

Doug Pitcock (Williams Brothers)

Geoff Rips (Rio Grande Legal Aid)

Amadeo Saenz (former executive director)

Thomas Sweeney (grandson of retired Highway Commission chairman Judge W. R. Ely)

Karen Threlkeld (public information officer, San Angelo District)

Darah Waldrip (Fleet Forward, TxDOT)

Charles Walker (formerly of Bridge Division; bridge historian and restoration engineer)

Anne Weisman (Hunter Industries)

John Weisman (Hunter Industries)

Dennis Wilde (former maintenance operations director, San Angelo District)

Martha Strain Wilkinson

Carol O'Keefe Wilson (author, *In the Governor's Shadow*)

Marcus Yancey (former deputy executive director)

Index

The letter *I* following a page number denotes an illustration, *m* denotes a map, and *p* denotes a photograph. Names of Texas State governors are followed by their dates in office.

Ball, Marcia, 292
Ball, Ted, 268p
Bankhead, John Hollis, 38
Bankhead Highway, 39, 167, 168p, 177
Bankhead Highway expedition, 59
Bankhead-Shackleford Act (Federal Aid
    Road Act), 39
Barnhart, Ray, 1, 287
Barrow, Clyde, 125, 148, 240–241, 241p
Barth, Brian, 309, 353p
Barth, Richard, 309
Barton, John, 293, 298p, 299, 340–341,
    345, 353p, 356p, 364, 366–367
Barton, Stephen Samuel, 27p
Bass, James, 355–356, 365–366, 366p
"Bats and Bridges" program, 314p
Battey, Shirley, 182p
Battle, Joe, 247, 249
Baytown Bridge, 254p
Baytown-East Freeway, 251
Baytown-La Porte Tunnel, 251, 252p,
    253, 253p, 255p
Bean, Roy, 160p
Beautification: Adopt-a-Highway
    program, 287–293, 290p; Don't Mess
    with Texas anti-litter campaign,
    288p, 290–293; roadside parks,
    137–147, 171p; tree preservation,
    135–137; wayside improvement and
    conservation, 135–147, 262, 264, 278
"Beautiful, Beautiful Texas" (O'Daniel),
    172
Bedier, Joseph, v
Behrens, Michael, 330p
Behrens, Mike, 325, 325p, 327, 330p,
    336–337, 342–343
Benson, Judy, 209p
Bergstrom, Bessie, 53, 90p, 117p
Bickett, John, Sr., 86–87, 95, 100
Bicycle Advisory Committee, 303
bicycles, 26–27, 26p, 34
Big Bend National Park, 368p
Big Bend Ranch State Park, 311p
Biggers, Don, 87
Bigham, Jack, 258
Biloxi, 18
Bjerke Management Solutions, 359
Black, A. C., 275
Black, Billy, 287–288
Black Bart (Charles Bowles), 239
Board of Control, 129
Board of Pardons, 101
Board of Water Engineers, 56

Bobbitt, Robert Lee, 209p
Boles, John, 156p
Bonnie and Clyde. *See* Barrow, Clyde;
    Parker, Bonnie
bootleggers, 239–240
Border Highway, 248
Bowles, Charles (aka Black Bart), 239
Bowser, O. P., 36
Brazos River Bridge, 91
Bribery and Corrupt Influence Statute,
    218
Bridge Division, 233, 324
bridges. *See also specific bridges*:
    automobile traffic on, 30p; "Bats
    and Bridges" program, 314p;
    big trucks impact on, 59; bond
    financing, 25; cable-stayed, 253,
    254p, 256p; construction standards,
    53; decorative elements, 326p;
    Ferguson era, 91–92; inspections,
    112p; number of, 67, 324; Oklahoma
    State coalition, 122–124; paint crews,
    120p, 152p, 316p; states connected
    by, 91–92, 111, 116, 122–124; trucks
    causing destruction on, 188p;
    underpasses, 132p, 134p, 176p;
    Briscoe, Dolph (1973–1979), 193, 217
Briscoe, John T., 36
Brown, David, 243
Buffalo Soldiers, 17
Bundesautobahnen, 213
Bureau of Intelligence, 148
Bureau of Public Roads, US, 39, 59, 61,
    244p
Bureau of Roadside Development, 135
Burka, Paul, 340
Burkett, Joe, 89, 95–96, 98
Burkham, Charles, 12
Burnett, Bill, 311–312, 312p, 315,
    317–318, 325p
Burnett, Richard, 312
Bush, George H. W., 304
Bush, George W. (1995–2000), 324, 335,
    340
Butler, John, 287
Butler, John R., Jr., 285p
Byrd, H. S., 176

Caddo, 18
Caddoan Mississippian Mound
    Builders, 9
Cage, John, 100
Calatrava, Santiago, 67p

217–218, 236, 272–273, 301, 341;
interstate highway system, 220–228,
230; military enlistments, 186–187,
190; post-WWII, 190, 192–193, 197;
safety focus, 184; small package
system, 149, 229, 334
Greer, Robert, 184, 186–187, 190
Greer Award, 270, 295
Guadalupe River Bridge, 209p
Gubbels, Jac, 144–147, 145p, 159, 185,
260, 264
Gulf Freeway (US 75), 198, 248, 250–251
Gulf Intercoastal Waterway, 282

Hagan, Hilton, 230, 230p
Hair, Mr., 128–129
Hall, Sam, 98p
Hamer, Frank, 125, 130p, 131, 148, 149p,
240
Hancock, Curtis, 45, 48, 53p, 56
Hank, R. J., 95, 117p
Hannaford, Alex, 242
Hansen, Anderson and Dunn, 250
Harbor Bridge, 233, 234p, 365
Harrigan, Mike, 241
Harris, Dilue Rose, 15
Harris, Roxanne, 328
Harrisburg Road, 30i
Harrison, Cora Sue, 307p
Hartman, Fred, 254
Hasinai, 9
Hayden-Cartwright Federal Highway
Act, 175
Heald, Charles Wesley "Wes," 309,
323–325, 323p, 325p, 327
Henderson, Elizabeth "little devil," 208p
Henry, Hubert, 268
Henry, Hubert A., 268p
Henry, Reeves, 33
High Five Interchange, 204p–205p, 298p
Highway 58, 146
Highway Act, 42–43
Highway Beautification Act, 260–263
Highway Department, Texas state: road
map, colonial Texas, 13
Highway Department, Texas state
(1917–1975). See also specific
administrators; specific departments;
administrative control, 42;
computerization, 264–265,
265p–266p, 267–269, 268p; contract
awards, 51, 149, 218–219, 229, 334;
debt, 114; demographics, 121;

Depression era, 120–121; district
engineers, importance to, 229; dress
code, 277, 277p; education and
training, 112; employee recognition
programs, 270; employee safety
program, 172; establishing the, 1,
36–37, 41–42; family culture, 130,
184, 273–274; federal aid; Allred
administration, 165; Dingwall
administration, 275; Duren
negotiations for, 51; Ferguson era,
86, 97, 125; Gilchrist administration,
114, 125, 149; Moody injunction,
97; requirements for, 38–39, 56,
61–62, 65, 105, 114, 165; World War
I, 55; federal regulations, delays
from, 275–276, 276p; growth, 273,
282; innovations, 233–235, 235p;
integrity, 111, 129–130, 162, 217–218,
236, 272–273, 301, 341; internal
auditing unit, 274–275; legislative
investigations, 129; Oklahoma State
coalition, 122–124; payroll, first, 52p;
post-WWII, 190, 192–193, 197; public
perception, 272–273; rebuilding,
105; responsibilities, early, 42–43,
51, 56; safety focus, 144–147, 184;
section foreman's car and mascot,
57p; structure, 42; testing standards,
230p; Texas A&M College alliance,
170; women employees, 170, 277;
World War II, 186–187, 190
highway departments, nation-wide
establishment of, 34
Highway Motor Patrol, 148, 148p
Highway Motor Pool, 114, 125, 135
The Highway of Hell, 242
Highway Patrol, 148
Highway Serial Killings Initiative, FBI,
241
highway system, nationally, 144
highway system, Texas state. See
also beautification; crime on the
highways, 239–243; displacement
and destruction, 198; first twenty-
six highways, 46, 166; maps. see
maps; road reform attempts, early,
36–37; safety in design, 114–116,
119p, 144–147, 184, 234–235, 235p;
truck damage to, 122, 236; youth
employment, 137, 144, 144p,
154–157, 221, 237–238, 238p
Highway Trust Fund, 275

Moore, S. B., 99–100
Morgan's Point Ferry, 251, 252p, 253
Moseley, Jeff, 281p
Motor Transport Convoy, 59, 61
Motor Transport Corps Convoy, 58–59, 175
Motor Vehicle Commission, 303
Motor Vehicle Division, 303
Motor Vehicles Division, 265
Munoz, Henry, III, 306
Murray, "Alfalfa Bill," 122–124
Murray, Jean, 124
"My Summer on I-45 (Norwood), 237–238, 238p

nail pickers, 206p–207p
Nalle, Ouida, 89
Natchitoch, 11
National Good Roads Association convention, 31
National Guard, 121–124
National Highway Program, President's Advisory Committee, 220–228
national highway system. See Interstate Highway System
National League for Good Roads, 27
National Youth Administration (NYA), 137, 144, 144p
Naylor, Harold, 100
Neches, 9
Neff, Pat M. (1921–1925), 62–63, 65, 77, 84, 85p, 87, 180
Nelson, Willie, ix, 143p, 292, 292p
New Deal, 144
New Hampshire, 276p
New Mexico, 246
newspapers, Ferguson's lawsuits against, 79–80
Nimitz, Chester William, 187
North Central Expressway, 198–205, 224
NorthPark Center, 225
Northside Community, 365
North Texas Tollway Authority, 340
Norwood, Larry, 237–238
NTT DATA, 358
Nuestra Señora Espíritu Santo de Zúñiga, 10

Obama, Barack, 345
O'Daniel, W. Lee "Pappy" (1939–1941), 172–173, 173p, 190
Odle, H. C., 45, 53p, 56
Oedipus, 239

Office of James Burnett and Jacobs Engineering Group, Inc., 359
Office of Road Inquiry, US, 25, 27
Ohio, 67, 324
oil boom, 121–122
oil crisis (1973), 280–281, 281p
oil crisis (1979), 281
oil industry, 50, 61, 67, 81–82, 187–188, 368
Oklahoma, 92, 116, 122, 240–241, 338
Old Age Assistance program, 164
Old Preston Road, 25
Old San Antonio Road (Camino Arriba), 9, 60p, 111m
Old Spanish Trail, 91p, 111
Oliver, Arnold, 297, 299–305, 302p, 308, 310–311, 325p
Oliver, Richard, 221, 221p
Olmec trails, 6
Olmos Construction, 333–334
Oregon, 135, 290
Our Sacred Monkeys (Biggers), 87

paint crews, 120p, 152p, 316p
Pakota (USS), 153
Pape, William, Sr., 137
parallel-log "corduroy" roads, 28
Parker, Bonnie, 125, 148, 240–241, 241p
Parks: roadside 137–147, 171p; state, 179–180, 244p; theme, 233, 233p
Parsons, C. N., 208p
pay-as-you-go policy, 217, 336, 366
Pease, Elisha, 18
Pecos River Bridge, 109p
pedestrian roads, 6
Pennybacker, Percy, 69p
Pennybacker, Percy V., 153
Pennybacker Bridge, 69p
Perry, Rick (2000–2015), 47, 209p, 324–325, 335–336, 340, 357
Pershing, John J., 38, 61
petroleum industry, 29. See also oil industry
Petry, Herbert C., Jr., 209p, 222–223, 224p
Phares, L. G., 117p
Pitcock, Doug, 218p, 218–219, 223
Plains bison, 2, 3m
Plan of San Diego, 38
Pointe aux Peconques (Pecan Point), 11–12
Pool, Walter, 12
Port Arthur-Orange Bridge, 153

Port of Corpus Christi, 365
postal service, 11
Post Highway, 208p
Post Road, 115p
Potts, R. J., 212p
Preston Trail, 25
Price, Betsy, 348
prisoner executions, 322
prisoner-of-war camps, 188
prisoner pardons, 93, 101, 103–104, 180
prisoners: convict labor, 31, 36, 43, 62p, 179–180, 181p; Greer supervision of, 179–180
prison labor, 181p
Prohibition, 239–240
Proposition One, 368–369
public highways: linking states, 39
Public Works Agency, 170

Queen Isabella Causeway, 296, 327–330, 327p, 329, 335, 343

railroad industry, 27–28, 32
railroads, 20–23, 22m
Rainbow Bridge, 70p, 150–153, 170, 316p
Rainy Day Fund, 368
Ralls, Mary Lou, 324p
Ramos, Basilio, Jr., 38
ranch-to-market roads, 193–196
Rand, Sally, 157
Rayburn, Sam, 221
Reagan, Ronald, 287
Reagan Houston, 183p
Reconstruction, 18–20
Recovery Act, 164
Red River Border War, 91p, 92, 122–124
Red River Bridge Company, 122
Regency Suspension Bridge, 68p
registration fees, 36
The Reivers (Faulkner), 29
rent-to-own system, 89–90
Republic of Texas, 16
Rhoades, Robert Ben, 242
Richards, Ann (1991–1995), 35, 305, 306p, 308, 313
Rimes, LeAnn, 292, 298p
Ringer, Harold, 316p
roadkill, 154–156
road management systems: nationwide, 34–36; resistance and controversies, 34–37
roads. See also construction of roads;

improvement of roads; maintenance of conquest and colonization, 3–4, 7–10, 214; countries connected by, 167, 338, 344; migration trails, 2–3, 7, 11, 19; national defense and, 58–59, 61, 175, 177–178, 177p, 187–188, 189p, 190, 213–214; paved, for "Good Roads" trains, 27–28; roadbeds, Caretas and carros drawn, 7; for rural communities, 25, 28; sacbeob, 4–6, 28; states connected by, 39, 55, 111m, 167, 246, 338; traffic increases on, 198, 251
roads, county, 36; funding, 39, 43; improvement, responsibility for, 62; inspections, 56; maintenance, 56, 65; responsibility for, 23–25, 31, 36, 43, 56
roads, Texas: 1919 statistics, 55
road tax, 18, 25
Robertson, Felix D., 84
Rogers, Ginger, 156p
Roosevelt, Franklin D., 133, 159, 164, 170, 175, 177
Rosa Parks Freeway, 290
Rosie (Hunter Industries accountant), 332–334
Route 66, 92–93, 92p
Runaway Scrape, 13, 15, 111
rural communities, 25, 28, 192–196
Rural Free Delivery (RFD), 27p

Sabine River bridge, 92
Sabine River Memorial Bridge, 111
Saburido, Jaqui, 293
sacbeob, 4–6, 28
Saenz, Amadeo, 328–329, 343–345, 344p–345p, 355–357, 367
safe driving, 184
safety: automobile accident, first, 33; "Click It or Ticket" seat belt campaign, 290–291; in highway design, 114–116, 119p, 144–147, 184, 234–235, 235p; traffic safety project, 219p
San Angelo District, 361–363
San Antonio de Valero (the Alamo). See Alamo (San Antonio de Valero)
San Antonio-El Paso Road, 17–18
Sanderford, Roy, 128
San Francisco de la Espada, 9
Sanitary Patrol, 154–156
San José de los Nazonis, 9